Praise from the Experts

"This exceptionally readable book will allow SQL programmers at all levels to enhance and expand their skills with this very versatile Base SAS procedure. Howard Schreier's approach of providing DATA step methods followed by SQL techniques to accomplish the same result is a very effective teaching technique, particularly for SAS programmers comfortable with the DATA step who want to add PROC SQL to their data management and reporting toolbox. The writing style is clear and concise, and the overall organization of the book is very logical and easy to follow. An exceptional feature of this book is the way that the explanation of each example provides insight into how PROC SQL 'thinks,' making the use of this procedure more intuitive for beginners and experienced users alike. In particular, the chapters on joins and subqueries, as well as the one on using the Macro Facility with PROC SQL, are wonderfully comprehensive and will provide a reference that you will turn to again and again."

Christianna Williams, PhD
Senior Associate, Abt Associates Inc.

"With a title like *PROC SQL by Example: Using SQL within SAS*, the author makes sure that the content of the text does exactly that for the reader. Howard Schreier has provided an abundance of clear, well-designed examples that detail the spectrum of SQL topics.

"The early chapters provide the reader with the appropriate SQL terminology with references to parallel Base SAS terms and concepts. Basic code and execution details are explained. Contrasts between SQL joins and DATA step merges are discussed at length. Most SQL examples are accompanied by companion DATA step or procedural examples. These comparisons are used to show the advantages of SQL capabilities over non-SQL SAS, as well as to identify any shortcomings of SQL where the non-SQL SAS solution might be more favorable. But the author points out that many solutions are equally efficient, whether programmed with SQL or Base SAS, so the solution chosen by the programmer might be based on personal coding preferences.

"Later chapters examine the advantages of using SQL for tasks such as creating macro variables, accessing and altering metadata, creating and using SAS views, and mixing SQL and non-SQL SAS code to complete a task. During the discussions on these features, the author has maintained easy-to-follow examples that often build upon previous examples presented in the text.

"Although this book is sold as a guide to SQL programming, SAS users might find themselves learning a few things about DATA step programming as well."

Stuart Long
Senior Systems Analyst, Westat

"This book has some great tips I will be putting to use! I've been programming with SAS and SQL for years, and this book is a really informative resource. I think learning by example is a thorough and efficient way to understand SAS techniques. The book has examples that you can pick up and use right away. I particularly enjoyed the comparisons between the Base SAS code and the PROC SQL code. It is a great bridge for SAS programmers to easily pick up this new language."

Candice Miller
Research Programmer, RAND Corporation

"*PROC SQL by Example: Using SQL within SAS* is an excellent introduction and reference manual for users looking to learn or brush up on their PROC SQL skills. The book does a great job of comparing SAS procedures to their PROC SQL counterparts. Using multiple examples, the book enables readers to easily follow along and grasp the knowledge needed to effectively enhance their SAS skills to tackle complex and new problems that might have been overwhelming before.

"Overall, *PROC SQL by Example* is a great introduction to the topic and will benefit those looking to make the jump forward."

Faisal Dosani
Senior Information Analyst

"*PROC SQL by Example: Using SQL within SAS* will appeal especially to those who have some experience with SAS DATA step programming and procedures, and who now have an interest in moving into SQL database programming. The author illustrates how SAS DATA steps and procedures translate into SQL syntax. More important, he also helps the reader understand how SQL programs embody logical constructs as opposed to procedural steps. Discussions of DATA step and SAS SQL programs build on actual programs that a reader can execute and modify. Perhaps for this reason, the text and examples achieve a level of precision and integrity that one rarely finds in a 'how to by example' text. Notable extra features include discussions of SAS SQL options, set operators, using PROC SQL as a report generator (as an alternative to PROC REPORT or PROC TABULATE), SAS MACRO scripting, 'pivoting' data structures using PROC TRANSPOSE, and data set persistence versus replacement.

"This book takes the programming cookbook genre to a new level of excellence."

Sigurd W. Hermansen
Westat

PROC SQL
by Example
Using SQL within SAS®

Howard Schreier

Contents

Chapter 7 **Global Statements, Options, and Session Management   131**

Chapter 8 **Using the Macro Facility with PROC SQL   141**

Chapter 9 **Table Maintenance and Alternate Strategies   161**

Chapter 10 Views   205

Acknowledgments

My name is the only one on the cover, but that doesn't mean I did it by myself. I thank the people at SAS Press and the reviewers whom they enlisted; all have been consistently helpful. I also thank those who taught me and, especially, those who encouraged me to teach myself.

Chapter 1

Introduction

SAS defines Structured Query Language (SQL) as "a standardized, widely used language that retrieves data from and updates data in tables and the views that are based on those tables" (see *Base SAS 9.2 Procedures Guide*: Procedures: The SQL Procedure: Overview). SQL is not an exclusive feature of SAS; it has been implemented by many vendors, and is especially widespread in the relational database management system (RDBMS) world. The SAS implementation of SQL is available in the SQL procedure (PROC SQL), part of Base SAS.

Some but not all PROC SQL capabilities are paralleled in the DATA step and in other SAS procedures. Thus, PROC SQL can be employed as a substitute for other elements of SAS or as a complement to those elements.

This book is intended for readers who are familiar with SAS but not with SQL, and who want to add PROC SQL to their SAS toolkits. It will also be useful to those familiar with other implementations of SQL who want to learn SAS.

1.1 More about SQL

PROC SQL is different from other SAS components and different from other software implementations of SQL.

Standards and Extensions

American National Standards Institute (ANSI) standard SQL is not a complete and self-sufficient language. For example, consider the definition quoted in the preceding section; it mentions retrieval and updating of data held in tables, but says nothing about how a table is populated in the first place. There are two possible approaches to the incompleteness. One is to include extensions (capabilities not required by the standard) in an SQL implementation to make the language more complete. Thus, for example, RDBMS vendors typically extend SQL with tools to import and export large volumes of data. The other approach, and the one followed by SAS, is to embed SQL into a language that provides the missing features. So, for example, a SAS application might use a PROC IMPORT step to load data, before turning to PROC SQL for processing and analysis of that data.

The implementation of SQL in SAS 9.2 PROC SQL does not fully comply with the current ANSI standard for SQL. On the other hand, PROC SQL includes some features **not** required by the standard.

Reference: For details about PROC SQL and the ANSI standard, see *Base SAS 9.2 Procedures Guide*: Procedures: The SQL Procedure: PROC SQL and the ANSI Standard.

Fundamental Differences between SQL and the DATA Step

The largest part of this book is devoted to explaining and illustrating the features of PROC SQL and identifying and qualifying parallels with non-SQL SAS counterparts to those features. Those explanations and examples deal with particular language elements. Before delving into that sort of detail, we should look at some general characteristics that distinguish PROC SQL from other parts of SAS. These distinctions range from the rather mundane to the almost profound.

Comma versus White Space Separation

In most parts of SAS, a series of like elements (such as variable names) is coded using white space (blanks, tabs, or new lines) for separation. In SQL, elements in such a series are separated by commas (with optional white space permitted in addition to, but not instead of, each comma).

Terminology

SAS and SQL both use two-dimensional collections of data, but have different terminology for the basic elements. The differences are summarized in the following table, which, for reference, extends to include the realms of data processing and relational database theory.

Realm	Corresponding Terms		
SAS	SAS data file	Observation	Variable
SQL	Table	Row	Column
Data processing	File	Record	Field
Relational database theory	Relation	Tuple	Attribute

The SQL and SAS terms are used more or less interchangeably throughout the book. We usually use SQL terms in discussing SQL code and SAS terms in discussing other code, but that's just a tendency and not a rule.

The term "SAS data file," though taken from the SAS documentation, might seem unfamiliar; perhaps you more often refer to a "SAS data set." The distinction is the inclusion of views. A SAS data set might be either a SAS data file or a SAS data view. In SQL, there is no similar umbrella term; there are tables and there are views.

In the remainder of the book, you will encounter many references to SAS data sets, and few if any to SAS data files. Meanings should be clear in context. In places where it's important to differentiate views from files (particularly in Chapter 10, which has views as its subject), more precise language is used.

Reference: For more information about SAS data files, see *SAS 9.2 Language Reference: Concepts*: SAS Files Concepts.

Executable Unit

In the DATA step, and in most (but not all) SAS procedures, the step is the executable unit of code. That is, SAS reads and interprets code from the beginning of a step and continues until it encounters another step boundary before it begins processing. In SQL, each **statement** is an executable unit. So SAS reads and interprets code until it encounters a semicolon (which is the statement terminator for SQL just as it is elsewhere in SAS) and then performs the requested processing before examining the next statement. The QUIT statement terminates PROC SQL (RUN statements are ignored).

Nestability

SQL code constructs can be nested; that is, a query can be an operand of another query (see Section 3.3 and Chapter 5). In the DATA step and other parts of SAS, such sequencing of computations can be implemented only by coding a chain of steps, with intermediate results typically passed forward as SAS data sets.

Namespace Management

A DATA step can accommodate only one variable of a given name. In PROC SQL, multiple columns having the same name can be used successfully if they are differentiated by prefixes or qualifiers indicating their sources (see Section 4.1 for details). Of course, when results are **stored** in SAS data sets, duplicate names are not allowed.

Procedural versus Declarative

When you use a **procedural** language, you tell the computer what to do, not what to produce. The SAS DATA step language is procedural, though that fact is sometimes masked by all of the defaults and automatic behavior. With a **declarative** language, such as SQL, you tell the computer what to produce. The translation of such specifications into an operational plan is the responsibility of the software and is largely hidden from the user.

Row Order

Even though rows of data are supplied to PROC SQL in some order, and even though PROC SQL output is stored or displayed in some order (which the programmer can specify), SQL conceptually treats a given table or view as an unordered set of rows. Consequently, a query cannot explicitly or implicitly make reference to row ordering.

Bias for Normalized Data Structure

The subject of data normalization is a big one, and well beyond the scope of this book. For our purposes, we consider a table having sets of similar columns requiring parallel treatment to be "denormalized." A table without such column sets, conversely, is termed "normalized," and characteristically has more rows and fewer columns than its denormalized counterpart. For example, suppose you have data on exports by country and year. If you store your data in a long table with just three columns (one to identify the country, one to indicate the year, and one to report the value of exports), you have a normalized table. On the other hand, if you use a matrix structure with one row for each country and one column to identify the country plus one column for each year's export data, your table is denormalized.

The DATA step is relatively neutral in supporting these alternative designs, with arrays and loops available to reference and process parallel columns. You could use an array and a loop to rather easily sum each year's exports. Many SAS procedures offer shortcuts for operating on sets of variables, so you could also use PROC MEANS with little difficulty to perform those aggregations. PROC SQL is different; it has nothing resembling arrays and loops. Consequently, SQL has a strong bias for normalized structures, and you will find SQL solutions much easier to develop if you organize your data accordingly. An example of such a normalized structure is presented in Section 12.3.

RDBMS Heritage

SQL comes to SAS from the world of relational database management systems, and that heritage shows in some ways. In particular, a lot of SQL statements are concerned with making changes to data **in place** (that is, inserting, deleting, or changing rows of data within a table, without replacing the table as a whole and without creating a new table coexistent with the original). SAS has such capabilities, but they are not widely used. SAS users are more likely to create new tables as they go along. We concentrate on doing things in that "SAS way" and confine our discussion of making changes in place to Chapter 9.

1.2 More about This Book

Before we take a detail-oriented look at PROC SQL in the following chapters, here are a few general how's and why's.

Purpose

The purpose of this book is to introduce PROC SQL to somewhat experienced SAS users. We start with the basics and then progress to more complex and specialized features.

We take almost every opportunity to illuminate SQL capabilities by demonstrating them together with more or less equivalent examples using the DATA step or SAS procedures other than PROC SQL. Because these non-SQL parallel techniques are not really the subject of the book, we don't explain them at length. However, we provide documentation references in those cases where the location of the relevant documentation might not be obvious, as well as references to *The Little SAS Book* (Fourth Edition). In addition, Chapter 14 provides something of a documentation "roadmap."

The SQL techniques presented in the book are specific to SAS in two ways. First, they occasionally use features of PROC SQL that extend beyond the requirements of the SQL standard. Second, they use language elements, such as SAS functions, that are not strictly part of SQL but rather are inherited from the SAS environment.

Perspective

Conceptually, and somewhat vaguely, we can divide the functionality of Base SAS software into four subsets. Moreover, we can visually suggest relationships among the four by positioning them in this diagram:

Beyond

Alternative   SQL

Shared

Since it is our subject, let's start with the quadrant identified as "SQL." It comprises the statements and options supported by PROC SQL. That's pretty clear-cut, and there is just one caveat, which is that not everything coded within a PROC SQL step belongs to this subset of Base SAS. For example, a formula within an SQL statement might include a reference to the ABS (absolute value) function, which is not part of PROC SQL per se, but rather is borrowed from the Base SAS function collection.

"Alternative" refers to capabilities that are equivalent (substantially if not completely) to PROC SQL functionality but are found in the DATA step and a handful of "workhorse" procedures (primarily PRINT, SORT, MEANS/SUMMARY, FREQ, DATASETS, and FORMAT). It's important to understand that this quadrant does not encompass **all** features of the DATA step and the enumerated procedures; rather, it comprises only those features having parallels in PROC SQL.

The "Shared" quadrant represents things that make SAS work, or work better, and that support both SQL and Alternative usage. This quadrant includes

- the user environment and interfaces (such as the Display Manager)
- libraries and engines (so that PROC SQL can access any data set that other parts of SAS can access)
- most functions (but not call routines)
- formats and informats
- data set options and SAS system options

- global statements (such as TITLE)
- the SAS Macro Facility
- the Output Delivery System
- some utility procedures, such as PROC IMPORT, PROC EXPORT, and parts of PROC DATASETS

"Beyond" comprises all features of Base SAS that are beyond the practical limitations of PROC SQL (and thus, by implication, beyond the boundaries of the Alternative quadrant). For example, consider PROC CHART, which produces low-resolution graphs. Such output is outside the capability of PROC SQL, so PROC CHART belongs in the Beyond quadrant.

This four-way partitioning of Base SAS is strictly a conceptual exercise intended to explain the purpose of this book. The subdivisions have no operational significance and in fact are not mentioned after this chapter. They are also, admittedly, a bit vague and arbitrary. For example, consider that PROC MEANS or PROC SUMMARY can calculate both means and medians, whereas PROC SQL can calculate only the means. The implication is that a PROC MEANS step that calculates both means and medians has one foot in the Alternative quadrant and one in the Beyond quadrant. Similarly, PROC DATASETS can be used to manage both integrity constraints and audit trails (see Section 9.7), but PROC SQL can manage only the former. Thus, a PROC DATASETS step that deals with both of these tools belongs to both the Alternative and Shared subsets.

Presentation

The first half of the book deals exclusively with data retrieval queries. Chapters 2 and 3 address simple queries (defined as those that draw data from a single source). The next three chapters (4, 5, and 6) cover queries that tap into multiple sources.

Later chapters address a number of relatively specialized or advanced topics. Chapter 7 deals with options and Chapter 8 with the Macro Facility. In Chapter 9, we move beyond data retrieval to explain SQL tools for changing data. Chapter 10 takes on the subject of views, and Chapter 11 addresses SQL features for generating reports and concludes the presentation of SQL features.

The final few chapters supplement the earlier material. Chapter 12 presents examples that emphasize the use of SQL as a complement to other parts of SAS. Chapter 13 provides a short introduction to the important issue of performance tuning. Finally, Chapter 14 is a bit of an essay on the SAS documentation of SQL.

The book as a whole covers the major features of PROC SQL with one significant exception: interoperation with third-party RDBMS products. PROC SQL includes the Pass-Through Facility, a mechanism for sending SQL code to other vendors' RDBMS products to be processed and then receiving the results for use by SAS. Pass-through requires availability (licensing and installation) of the appropriate SAS/ACCESS product. The passed-through code must be written in the SQL "dialect" of the target RDBMS. Thus, working examples would depend on the choice of target system. For that reason, and because the focus of this book is on the Base SAS context, we do not discuss the Pass-Through Facility. SAS/ACCESS also permits SAS to operate with RDBMS servers more transparently, via LIBNAME statements. The behavior of PROC SQL code that exploits this feature depends on the nature of such code and the capabilities of the particular target RDBMS, and is also beyond the scope of the book.

Reference: For more information about PROC SQL interoperation with third-party RDBMS products, see *Base SAS 9.2 Procedures Guide*: Procedures: The SQL Procedure: Concepts: SQL Procedure: What Is the Pass-Through Facility? and Connecting to a DBMS Using the LIBNAME Statement.

This book does not attempt to repackage all of the information provided in the SAS documentation. For example, the use of PROC SQL options is explained (see Section 7.2), and a number of those options are identified and described there and elsewhere, but the coverage of such options is not comprehensive. The SAS documentation is available in four forms:

- hard copy
- online (Web pages at support.sas.com)
- PDF (Portable Document Format) files, available at support.sas.com
- locally installed Help files

All but the first are essentially free of cost. This book supports use of the documentation by providing numerous references. In addition, Chapter 14 is devoted to a discussion of the PROC SQL documentation.

The examples presented in the book were all set up and run using SAS 9.2 running on a Microsoft Windows XP host system. With the exception of a handful of LIBNAME statements, none of the code is specific to a host system, so it should be possible to run the examples on any SAS 9.2 platform and get the results shown. Because the PROC SQL feature set changed little between SAS 9.1.3 and SAS 9.2, there should be little difficulty replicating results with SAS 9.1.3.

All data libraries in the examples use the default native engine. The extent to which other engines can be successfully substituted will vary. Generally, there should not be problems in **reading** data from any engine. However, not all engines support all features for output.

The results displayed in the book were generated by running the examples in batch mode. However, there should be little or no effect on the results if Display Manager or other SAS user interfaces are used instead.

The examples attempt to simplify aspects of the code that are incidental to the subject at hand. So, for instance, nearly all tables in the examples are stored in the WORK library and denoted with one-part names (not prefixed with library references).

Reference: Read more about SAS libraries and library references in *SAS 9.2 Language Reference: Concepts*: SAS Files Concepts: SAS Data Libraries and in Section 1.11 of *The Little SAS Book* (Fourth Edition).

Now let's look at some of the visual properties of the examples. To illustrate, we'll borrow some bits and pieces from later chapters. Don't worry now about understanding the substance; explanations are provided in the original contexts.

First, here is a code specimen:

```
PROC SQL;
SELECT      fname, age
FROM        preteen;
QUIT;
```

Notice that it is indented and immediately follows a colon in the preceding text.

The use of uppercase and lowercase letters is significant in the examples. In code, names of SAS **language** elements (statements, clauses, options, functions, formats, and so on) appear in uppercase. The names of user-specified **data** elements (library references, data sets, variables, and so on) appear in lowercase or mixed case. This distinction applies only to code paragraphs (distinguishable by indention and use of a monospace font). In narrative paragraphs, **both** language and data element names are in uppercase; for example, we might refer to the FNAME column.

Notice that the preceding example includes a PROC SQL statement to launch the procedure and a QUIT statement to close the procedure. These statements are included because the SQL statements (such as the SELECT statement) work only if PROC SQL is already running. However, you should not interpret this to mean that each SQL statement must necessarily be in a separate procedure step. See Chapter 7 for more about PROC SQL session management.

The coded examples generate three types of results (although any particular example might involve only one or two):

- text appearing in the SAS log

- displayed output (that is, material that appears in the Output window—assuming that the code was run via the Display Manager)
- data sets produced or changed by the code

Let's look at these, starting with some log text:

```
NOTE: Table WORK.NEW created, with 7 rows and 5 columns.
```

The log text is set off with a ruled box rather than by indention. Now here is some displayed output:

```
 Median
   of 1
--------
    1.1
      6
    7.7
```

The presentation box is just like that used for log text. Distinctions between the two are either obvious or are noted in the narrative.

Examples in this book were run with the following SAS system options in effect: FORMCHAR="|----|+|---+=|-∧<>*" NOCENTER NODATE NOOVP.

Finally, let's see a result in the form of a data set. Of course, we are less interested in the hard-to-read bits and bytes of the data set as it is stored on disk than in a rendering of its content. Such a rendering is in Exhibit 1-1. Notice that this rendering is titled as an exhibit.

Exhibit 1-1 A table borrowed from a subsequent chapter

Sex	Youngest	Oldest	Avg_Height	Avg_Weight
F	11	12	55.8	70.7
M	11	12	59.7	98.9

Chapter 2

Basic Building Blocks

We begin exploring SQL by looking at the SELECT statement, which is the core of the language. The emphasis for now is on what can be called "building blocks." These are basic capabilities that, for the sake of simplicity, we examine in isolation. Later we'll see how they can be combined. In this chapter we look at the most basic of these building blocks, those that are equivalent to single non-SQL SAS steps.

The SELECT statement can stand on its own, and we start by seeing it in that context, although it is more often used as a clause within another statement. Essentially, the SELECT statement tells the SQL processor to extract, and possibly further process, a rectangular body of data comprising one or more columns and zero or more rows.

Before looking at any SQL code, let's create a table to be the basis of the examples in this chapter (most of them, anyway). To make things compact, we construct a subset of the SASHELP.CLASS data set:

```
DATA preteen;
SET sashelp.class;
WHERE age<13;
LABEL  name = 'First Name';
RENAME name = FName;
FORMAT height weight 5.1;
RUN;
```

The result is shown in Exhibit 2-1.

Exhibit 2-1 PRETEEN

FName	Sex	Age	Height	Weight
James	M	12	57.3	83.0
Jane	F	12	59.8	84.5
John	M	12	59.0	99.5
Joyce	F	11	51.3	50.5
Louise	F	12	56.3	77.0
Robert	M	12	64.8	128.0
Thomas	M	11	57.5	85.0

2.1 The Simplest SELECT Statement

We start with what is arguably the most primitive possible SELECT statement, in that it contains only the mandatory elements and utilizes simple forms for these elements. A SELECT statement must include a FROM clause designating the source of the data, preceded by a list of columns. So here is our extremely basic SELECT statement:

```
PROC SQL;
SELECT      *
FROM        preteen
;
QUIT;
```

PRETEEN is the name of the source table, and the asterisk (*) is a shorthand form telling the SQL processor to take **all** columns found in the source.

What does the processor then do with the columns? It displays them. The output looks like this:

```
First
Name      Sex      Age  Height  Weight
--------------------------------------
James     M         12    57.3    83.0
Jane      F         12    59.8    84.5
John      M         12    59.0    99.5
Joyce     F         11    51.3    50.5
Louise    F         12    56.3    77.0
Robert    M         12    64.8   128.0
Thomas    M         11    57.5    85.0
```

We've said that PROC SQL displays this little report, but we haven't said where. It depends on how SAS is being used and on how it is configured. If you are using the Display Manager with defaults in effect, you will see it in the Output window.

Preview: For now, we are using simple tables or views as sources. Later we will see more elaborate FROM clauses, which can preprocess the data for the SELECT statement (see the discussion of inline views in Section 3.3) or integrate data from multiple tables or views (see Chapter 4 on joins).

So this simple SELECT statement is more or less equivalent to the following PROC PRINT step:

```
PROC PRINT DATA=preteen;
RUN;
```

which generates:

```
Obs    FName     Sex    Age    Height    Weight

 1     James     M      12      57.3      83.0
 2     Jane      F      12      59.8      84.5
 3     John      M      12      59.0      99.5
 4     Joyce     F      11      51.3      50.5
 5     Louise    F      12      56.3      77.0
 6     Robert    M      12      64.8     128.0
 7     Thomas    M      11      57.5      85.0
```

Comparing this result with the output from the SQL SELECT statement, we notice the presence of a reference column of observation numbers and the absence of variable labels (the second column is headed "FName," in contrast to the SQL output, which uses the variable's label, "First Name"). So the SELECT statement more closely resembles PROC PRINT with the NOOBS and LABEL options in effect, as in:

```
PROC PRINT NOOBS LABEL DATA=preteen;
RUN;
```

which yields:

```
First
Name     Sex    Age    Height    Weight

James    M      12      57.3      83.0
Jane     F      12      59.8      84.5
John     M      12      59.0      99.5
Joyce    F      11      51.3      50.5
Louise   F      12      56.3      77.0
Robert   M      12      64.8     128.0
Thomas   M      11      57.5      85.0
```

Reference: Read more about PROC PRINT in the *Base SAS 9.2 Procedures Guide*: Procedures: The PRINT Procedure or in *The Little SAS Book* (Fourth Edition): Section 4.4.

Still, there are minor differences in layout. Most obvious is the horizontal line that PROC SQL places between the headings and the values. Also, the PROC PRINT column headings are centered, whereas headings in the SQL output are left-justified (for character columns) or right-justified (for numeric columns).

Preview: Later, we'll see how PROC SQL can be made to provide reference numbers like those in PROC PRINT's "Obs" column (see Chapter 11).

Actually, this example is a bit contrived in one way. Remember that in PRETEEN we specified the 5.1 format for the numeric variables HEIGHT and WEIGHT. Both PROC SQL and PROC PRINT then used that format; that's why we've seen one decimal place in all of the HEIGHT and WEIGHT values. However, in the absence of such format declarations, PROC SQL and PROC PRINT employ somewhat different rules for internally determining the number of decimal places to present.

2.2 A More Selective SELECT

What if we don't want to display all of the columns? You might recall that PROC PRINT lets you include a VAR statement to enumerate the variables, like this:

```
PROC PRINT NOOBS LABEL DATA=preteen;
VAR fname age;
RUN;
```

The result looks like this:

```
First
 Name      Age

James       12
Jane        12
John        12
Joyce       11
Louise      12
Robert      12
Thomas      11
```

The equivalent in SQL is to replace the shorthand asterisk in the SELECT statement with a comma-separated list of the columns to be included, like this:

```
PROC SQL;
SELECT       fname, age
FROM         preteen
;
QUIT;
```

The result is:

```
First
Name           Age
------------------
James           12
Jane            12
John            12
Joyce           11
Louise          12
Robert          12
Thomas          11
```

2.3 Storing Results

Very often you don't want to display results. Instead you want to store them for use in subsequent computations. That's what this DATA step will do:

```
DATA new;
SET preteen;
RUN;
```

The log confirms that a data set has been stored:

```
NOTE: The data set WORK.NEW has 7 observations and 5
variables.
```

Of course this is a rather trivial example in that it makes what is essentially an exact copy of the source. Its purpose is just to illustrate the mechanics of creating a data set to contain results. To do the same thing with PROC SQL, we use the CREATE statement:

```
PROC SQL;
CREATE TABLE new AS
SELECT      *
FROM        preteen
;
QUIT;
```

The SELECT statement, which we used earlier to display data, is now subordinated as a clause within the CREATE statement. Nothing is displayed; instead the columns of the existing table (PRETEEN) are stored in a new table. The log reports:

```
NOTE: Table WORK.NEW created, with 7 rows and 5 columns.
```

Notice that this note uses SQL terminology ("table," "rows," and "columns") rather than the usual SAS terminology ("data set," "observations," and "variables"). Nevertheless, it's the same message.

2.4 Column Subsets

What if you don't need all of the variables available in the existing data set? In the DATA step, a KEEP statement can be used to identify those to be stored in the new data set. For example:

```
DATA subset;
SET preteen;
KEEP fname sex age;
RUN;
```

We've already seen an analogous construct in PROC SQL, when we restricted the columns to be displayed in order to emulate the behavior of the VAR statement in PROC PRINT (see Section 2.2). We did that by coding a comma-separated list enumerating individual columns. It worked in a freestanding SELECT statement, and it also works within a CREATE statement. So the code is:

```
PROC SQL;
CREATE TABLE subset AS
SELECT      fname, sex, age
FROM        preteen
;
QUIT;
```

The new table, SUBSET, is as shown in Exhibit 2-2, whether it's produced with a DATA step or with SQL.

Exhibit 2-2 SUBSET

FName	Sex	Age
James	M	12
Jane	F	12
John	M	12
Joyce	F	11
Louise	F	12
Robert	M	12
Thomas	M	11

PROC SQL has no counterpart to the DROP statement, which is sort of the mirror image of the KEEP statement, enumerating variables that are **not** to be placed in the output data set, as in:

```
DATA subset;
SET preteen;
DROP height weight;
RUN;
```

However, this approach to subsetting can be implemented if we turn to SAS features available within PROC SQL. Specifically, we can code the DROP= data set option for the table being created, as in:

```
PROC SQL;
CREATE TABLE subset(DROP=height weight) AS
SELECT      *
FROM        preteen
;
QUIT;
```

Tip: Most SAS data set options can be used in PROC SQL.

2.5 New Columns

It's often necessary to create a new variable using a formula. In the DATA step, that's done with an assignment statement, like this:

```
DATA ratios;
SET preteen;
ATTRIB Ratio FORMAT=5.2 LABEL='Weight:Height Ratio';
ratio = weight / height;
RUN;
```

The ATTRIB statement here is not essential. It just provides a variable label and an appropriate format.

The equivalent in SQL is to insert the formula in the column selection list, followed by the keyword AS and the name. To continue with the example, the SQL statement would be:

```
PROC SQL;
CREATE TABLE ratios AS
SELECT       *,
             weight / height AS Ratio
              FORMAT=5.2 LABEL='Weight:Height Ratio'
FROM         preteen
;
QUIT;
```

The selection list begins with the asterisk (*), which is an abbreviated way of calling for all of the columns in the source table (PRETEEN). It then continues with the formula for our new column. The FORMAT and LABEL specifications serve the same purpose as the ATTRIB statement in the DATA step and, like the ATTRIB statement, are not absolutely necessary. In fact, even the AS clause providing the name for the new column could be omitted; an automatically generated (but not very meaningful) name would appear instead.

The result, whether created in a DATA step or by PROC SQL, looks like Exhibit 2-3.

Exhibit 2-3 RATIOS

FName	Sex	Age	Height	Weight	Ratio
James	M	12	57.3	83.0	1.45
Jane	F	12	59.8	84.5	1.41
John	M	12	59.0	99.5	1.69
Joyce	F	11	51.3	50.5	0.98
Louise	F	12	56.3	77.0	1.37
Robert	M	12	64.8	128.0	1.98
Thomas	M	11	57.5	85.0	1.48

2.6 Aggregation

We often need to derive summary statistics from our data. SAS provides a variety of methods for doing this. One of the most versatile is PROC SUMMARY. SQL does not have nearly the extent of functionality provided by a specialized tool like PROC SUMMARY, but it is an alternative for a lot of relatively simple tasks.

Reference: Read more about PROC SUMMARY and its sibling PROC MEANS in the SAS documentation and in Sections 4.9 and 4.10 of *The Little SAS Book* (Fourth Edition).

Grand Totals and More

Here is a simple example in which PROC SUMMARY is used to produce one row of aggregate measures for our PRETEEN table:

```
PROC SUMMARY DATA=preteen;
VAR age height weight;
OUTPUT OUT=overall_averages(DROP = _type_ _freq_)
 MIN (age   )=Youngest
 MAX (age   )=Oldest
 MEAN(height)=Avg_Height
 MEAN(weight)=Avg_Weight;
RUN;
```

The procedure was instructed to compute the extrema of AGE and the averages of HEIGHT and WEIGHT. The result is shown in Exhibit 2-4.

Exhibit 2-4 OVERALL_AVERAGES

Youngest	Oldest	Avg_Height	Avg_Weight
11	12	58.0	86.8

Notice that the averages derived from HEIGHT and WEIGHT are displayed with exactly one decimal place. That's because averages computed by PROC SUMMARY inherit the FORMAT attributes from the underlying variables.

We can do the same thing in PROC SQL by coding a column selection for each statistic to be produced, like this:

```
PROC SQL;
CREATE TABLE overall_averages AS
SELECT      MIN (age)     AS Youngest,
            MAX (age)     AS Oldest,
            MEAN(height)  AS Avg_Height FORMAT=5.1,
            MEAN(weight)  AS Avg_Weight FORMAT=5.1
FROM        preteen
;
QUIT;
```

AS clauses follow each column selection to provide names. We include FORMAT specifications because in PROC SQL they are **not** inherited from the columns being aggregated. The output table is indistinguishable from the one produced by PROC SUMMARY.

Subtotals and More

Now suppose that instead of an overall summary, we want the computations stratified by SEX. The PROC SUMMARY code shown previously can be adapted by inserting a CLASS statement and coding the NWAY option (to suppress production of the grand overall statistics, which we no longer want). Here is the code:

```
PROC SUMMARY DATA=preteen NWAY;
CLASS sex;
VAR age height weight;
OUTPUT OUT=group_averages(DROP = _type_ _freq_)
 MIN (age    )=Youngest
 MAX (age    )=Oldest
 MEAN(height)=Avg_Height
 MEAN(weight)=Avg_Weight;
RUN;
```

The result appears in Exhibit 2-5. Notice that **PROC SUMMARY** automatically includes CLASS variables in output data sets.

Exhibit 2-5 GROUP_AVERAGES

Sex	Youngest	Oldest	Avg_Height	Avg_Weight
F	11	12	55.8	70.7
M	11	12	59.7	98.9

To accomplish the same thing using **PROC SQL**, we adapt the code presented in the previous section by inserting a GROUP BY clause:

```
PROC SQL;
CREATE TABLE group_averages AS
SELECT      sex,
            MIN (age)    AS Youngest,
            MAX (age)    AS Oldest,
            MEAN(height) AS Avg_Height FORMAT=5.1,
            MEAN(weight) AS Avg_Weight FORMAT=5.1
FROM        preteen
GROUP BY    sex
;
QUIT;
```

We also include the column SEX in the selection list. If we don't, the computations would be unaffected, but the two rows of results would not be identified, because PROC SQL does **not** automatically include GROUP BY columns. The results from PROC SQL are identical to those from PROC SUMMARY (see Exhibit 2-5).

PROC SQL is strict about the order in which clauses are coded. The GROUP BY clause must appear after the FROM clause.

Details

It's important to recognize the distinction between the SQL-specific summary statistic functions (like MIN) used in the preceding examples in this chapter and the corresponding SAS functions that can be used in PROC SQL as well as in the DATA step and other contexts. The syntax for their use is the same, and the repertoires of statistics (mean, maximum, minimum, and so on) are similar, but they are different tools.

The SQL-only functions are akin to the keywords used in PROC SUMMARY. Those keywords call for **vertical** aggregation. That is, processing is done separately for each variable, drawing values from as many observations as there are.

SAS functions, as used in DATA step code, operate **horizontally**. They can, and usually do, have multiple arguments. A value is drawn from each argument, using data currently available (and thus typically originating in a single observation).

In PROC SQL, both families of functions can be used. This creates ambiguity in some situations. Moreover, not all of the statistics available as SAS functions are available as SQL-only functions, and some keywords supported in PROC SUMMARY are not supported as PROC SQL functions. All in all, it's a situation that can be a bit precarious.

Reference: Read more about these functions in the *SAS 9.2 Language Reference: Dictionary*: Dictionary of Language Elements: Functions and CALL Routines: Functions and CALL Routines by Category: Descriptive Statistics and in the *Base SAS 9.2 Procedures Guide*: Appendices: SAS Elementary Statistics Procedures: Keywords and Formulas.

An example should help. Let's start with a little 3x3 table built as follows:

```
DATA threex3;
INPUT a b c;
CARDS;
1.1 2.0 3.0
6.0 5.0 4.4
7.7 8.0 9.0
;
```

We want to derive means (averages) and medians. When there are two or more arguments, there is no ambiguity, because a vertical function is restricted to a single argument. So the following code calls the SAS (**not** SQL-specific) MEAN function to average the three values in each row:

```
PROC SQL;
SELECT      MEAN(a,b,c) LABEL='Mean of 3'
FROM        threex3
;
QUIT;
```

The result is:

```
    Mean
    of 3
 --------
2.033333
5.133333
8.233333
```

Similarly, we can use the **MEDIAN** function:

```
PROC SQL;
SELECT      MEDIAN(a,b,c) LABEL='Median of 3'
FROM        threex3
;
QUIT;
```

This is the result:

```
  Median
    of 3
 --------
       2
       5
       8
```

Again, the calculations are confined, row by row. With two arguments, that doesn't change. But let's see what happens when we cut back to a single argument, as in:

```
PROC SQL;
SELECT      MEAN(a) LABEL='Mean of 1'
FROM        threex3
;
QUIT;
```

Our output is:

```
     Mean
     of 1
 --------
4.933333
```

This is the mean of the values in column A, calculated using all of the rows. The syntax itself, MEAN(A), is ambiguous. It conceivably could be interpreted as a horizontal function call requesting the mean of a single value (admittedly a triviality) within each row. However, PROC SQL has an internal decision rule to resolve the ambiguity. Whenever a construct of this form (a summary statistic name followed by a single argument in parentheses) could refer to either a SAS function (horizontal) or an SQL function (vertical), it is assumed to refer to the latter. PROC SQL then computes it vertically.

Now let's try it with **MEDIAN**:

```
PROC SQL;
SELECT      MEDIAN(a) LABEL='Median of 1'
FROM        threex3
;
QUIT;
```

We get:

```
  Median
    of 1
 --------
     1.1
       6
     7.7
```

You were probably expecting to see just a 6, the median of the values in column A. Instead we have, for each row, the trivial median of a single value of A. In other words, the processing was horizontal rather than vertical.

The explanation: vertical calculation of medians is not supported in PROC SQL (though it is in PROC SUMMARY). Thus, there is no ambiguity. In SQL, the only valid interpretation of **MEDIAN** with a single argument is that it is a SAS function call, to be computed horizontally.

Perhaps it would be better if PROC SQL instead raised an error in this situation, but it doesn't. So, you, the programmer, have to understand the decision rules and be informed about just which statistics are supported.

> **Tip:** PROC SQL supports the operation of four functions (MIN, MAX, N, and NMISS) with character data. This usage is limited to vertical summarization of a single argument.

2.7 Conditionality

It is not uncommon to have values that depend on other values—in other words, conditionality. Probably the most common way of implementing conditionality in the DATA step is the IF/THEN/ELSE structure.

For example, suppose that students of different ages and sexes are to go on different field trips. The 11-year-olds (boys and girls) are going to the zoo; girls who are not going to the zoo (that is, 12-year-old girls) are going to the museum; and boys who aren't going to the zoo have to stay behind. Here's one way of generating a list of individual student destinations:

```
DATA trip_list;
SET preteen;
IF      age=11  THEN Trip = 'Zoo    ';
ELSE IF sex='F' THEN trip = 'Museum';
ELSE                 trip = '[None]';
KEEP fname age sex trip;
RUN;
```

The result appears in Exhibit 2-6.

Exhibit 2-6 TRIP_LIST

FName	Sex	Age	Trip
James	M	12	[None]
Jane	F	12	Museum
John	M	12	[None]
Joyce	F	11	Zoo
Louise	F	12	Museum
Robert	M	12	[None]
Thomas	M	11	Zoo

Before looking for the SQL equivalent, we should probably note that, in an important sense, there is no equivalent. That's because the IF/THEN/ELSE combination triggers the execution of statements, which in turn perform actions. That is very much a procedural construct and thus alien to SQL, which is a nonprocedural language.

However, in this example the conditional statements are assignment statements, which evaluate expressions and store the results. Moreover, they all have the same target variable (TRIP). In this special case, SQL **does** have a near counterpart, the CASE structure. So here is the SQL effective equivalent of the DATA step that builds the table TRIP_LIST:

```
PROC SQL;
CREATE TABLE trip_list AS
SELECT       fname,
             age,
             sex,
             CASE WHEN age=11   THEN 'Zoo'
                  WHEN sex='F'  THEN 'Museum'
                  ELSE              '[None]'
                  END
             AS Trip
FROM         preteen
;
QUIT;
```

This again produces the result shown in Exhibit 2-6.

The CASE structure begins with the keyword CASE and ends with END. Within, WHEN/THEN specifications are examined in order until an expression following a WHEN is evaluated as TRUE. When that occurs, the expression following the THEN is evaluated to provide the result. If no WHEN condition holds true, the ELSE specification provides the result. Note that results can depend on the order in which the WHEN/THEN specifications appear.

If you are familiar with the DATA step SELECT statement (not to be confused with the SQL SELECT statement), the CASE expression probably looks a bit familiar. The two have many parallels in design. Here is the DATA step solution implemented with SELECT:

```
DATA trip_list;
SET preteen;
SELECT;
   WHEN (age=11)  Trip = 'Zoo   ';
   WHEN (sex='F') trip = 'Museum';
   OTHERWISE      trip = '[None]';
   END;
KEEP fname age sex trip;
RUN;
```

2.8 Filtering

Earlier in this chapter, we saw how column-oriented subsetting can be accomplished by enumerating the columns being selected (rather than using the inclusive asterisk notation). Now we turn to row-oriented subsetting.

Filtering from the Source

In the DATA step, observations (rows) to be processed can be included or excluded by means of the WHERE statement. To illustrate:

```
DATA girls;
SET preteen;
WHERE sex='F';
RUN;
```

yields Exhibit 2-7.

Exhibit 2-7 GIRLS

Name	Sex	Age	Height	Weight
Jane	F	12	59.8	84.5
Joyce	F	11	51.3	50.5
Louise	F	12	56.3	77.0

WHERE processing is available not only in the DATA step, but also in most SAS procedures, including PROC SQL. Translation to SQL requires no change other than to convert the WHERE statement to a clause within the SELECT clause or statement. So we can produce the same table by running:

```
PROC SQL;
CREATE TABLE girls AS
SELECT      *
FROM        preteen
WHERE       sex='F'
;
QUIT;
```

In both environments, the filtering is done "up front," by the engine that reads the data.

The order of the clauses under the SELECT statement or clause is dictated. The WHERE clause follows the FROM clause and precedes the GROUP BY clause, if there is one.

Now let's see what happens when the filter eliminates **all** of the data. We can do that first with a stand-alone SELECT statement such as this:

```
PROC SQL;
SELECT      *
FROM        preteen
WHERE       age=10
;
QUIT;
```

When we run this code, no output is produced, but in the log we see:

```
NOTE: No rows were selected.
```

When we embed the SELECT in a CREATE TABLE statement, the table is created, but it contains no rows:

```
PROC SQL;
CREATE TABLE tens AS
SELECT      *
FROM        preteen
WHERE       age=10
;
QUIT;
```

In the log we see just the standard note:

```
NOTE: Table WORK.TENS created, with 0 rows and 5
columns.
```

Filtering Aggregated Data

SQL has a second filtering device, the HAVING clause. The distinction between this and the WHERE clause is that HAVING conditions can reference summary statistics and are evaluated **after** aggregations are performed. Thus they take effect "downstream," on the output side of the process.

To illustrate, consider this PROC SUMMARY step, which calculates the extreme values of the HEIGHT variable and does so separately for each SEX/AGE combination:

```
PROC SUMMARY DATA=preteen NWAY;
CLASS sex age;
OUTPUT MAX(height)=Tallest MIN(height)=Shortest
 OUT= hilo(DROP = _type_ _freq_);
RUN;
```

The result is presented in Exhibit 2-8.

Exhibit 2-8 HILO (unfiltered)

Sex	Age	Tallest	Shortest
F	11	51.3	51.3
F	12	59.8	56.3
M	11	57.5	57.5
M	12	64.8	57.3

The equivalent SQL code is:

```
PROC SQL;
CREATE TABLE hilo AS
SELECT     sex,
           age,
           MAX(height) AS Tallest,
           MIN(height) AS Shortest
FROM       preteen
GROUP BY   sex, age
;
QUIT;
```

There is nothing here we have not encountered when we looked at aggregation techniques earlier in this chapter. But now suppose we want to limit the results to those SEX/AGE combinations where the difference between the height extrema is four or more inches. In PROC SUMMARY, we could insert a WHERE= data set option to accomplish this. The code would become:

```
PROC SUMMARY DATA=preteen NWAY;
CLASS sex age;
OUTPUT MAX(height)=Tallest MIN(height)=Shortest
 OUT=hilo(WHERE = (tallest - shortest > 4)
 DROP = _type_ _freq_ );
RUN;
```

The result can be seen in Exhibit 2-9.

Exhibit 2-9 HILO (filtered)

Sex	Age	Tallest	Shortest
M	12	64.8	57.3

PROC SUMMARY still computed the minima and maxima for all four SEX/AGE combinations, but the WHERE= condition prevented three of the four from being written to the output data set because the difference between minimum and maximum did not exceed four inches.

We could of course use the same WHERE= data set option in our SQL code. However, SQL has a device intended precisely for this situation: the HAVING clause. So we'll use that rather than borrowing a SAS option.

Once again, clause order is important. The HAVING clause must follow the FROM clause and the GROUP BY clause, if there is one. So the modified SQL code is:

```
PROC SQL;
CREATE TABLE hilo AS
SELECT       sex,
             age,
             MAX(height) AS Tallest,
             MIN(height) AS Shortest
FROM         preteen
GROUP BY     sex, age
HAVING       tallest - shortest > 4
;
QUIT;
```

It again produces the results in Exhibit 2-9.

2.9 Reordering Rows

The purpose of **PROC SORT** is the reordering of observations. For example, if we run:

```
PROC SORT DATA=preteen OUT=age_sort;
BY DESCENDING age fname;
RUN;
```

the newly created data set looks like Exhibit 2-10.

Exhibit 2-10 AGE_SORT

FName	Sex	Age	Height	Weight
James	M	12	57.3	83.0
Jane	F	12	59.8	84.5
John	M	12	59.0	99.5
Louise	F	12	56.3	77.0
Robert	M	12	64.8	128.0
Joyce	F	11	51.3	50.5
Thomas	M	11	57.5	85.0

The 12-year-olds appear before the 11-year-olds because of the DESCENDING keyword; the names are alphabetical within each age group.

PROC SQL has equivalent capabilities. The SQL code equivalent to our PROC SORT step is:

```
PROC SQL;
CREATE TABLE age_sort AS
SELECT     *
FROM       preteen
ORDER BY   age DESCENDING, fname
;
QUIT;
```

Notice that the keyword DESCENDING follows rather than precedes the name of the column to which it pertains.

As explained earlier, row order is conceptually irrelevant to the specification of queries using SQL (see Section 1.1). Of course, when results are displayed or stored, you might need to have them in a particular order for your purposes. That's why the ORDER BY clause is part of the language. However, operation of the ORDER BY clause should be thought of as a sort of postprocessing, done **after** conceptually unordered results are derived, but **before** those results are displayed or stored. For this reason, when an ORDER BY clause is used, it's almost always the last clause in a statement, immediately preceding the semicolon (a rarely encountered exception involves the USING clause; see Section 10.5).

2.10 Elimination of Duplicates

Eliminating duplicate rows from a table is a common task. To illustrate, we first need to have a data set containing duplicates. We'll get one by eliminating some of the columns in PRETEEN:

```
CREATE TABLE sex_age AS
SELECT sex, age
FROM   preteen
;
```

The result, which we've named SEX_AGE, is presented in Exhibit 2-11.

Exhibit 2-11 SEX_AGE

Sex	Age
M	12
F	12
M	12
F	11
F	12
M	12
M	11

A commonly used technique for elimination of the duplicates is to use the NODUPRECS option of PROC SORT:

```
PROC SORT DATA=sex_age OUT=sex_age_distinct NODUPRECS;
BY _ALL_;
RUN;
```

Note the use of a special SAS name list (_ALL_) to include all variables as sort keys. This assures that identical observations will be gathered consecutively during the sorting process, which in turn assures that all duplicates will be eliminated. So the result is as presented in Exhibit 2-12.

Exhibit 2-12 SEX_AGE_DISTINCT

Sex	Age
F	11
F	12
M	11
M	12

Reference: Read more about special SAS name lists in *SAS 9.2 Language Reference: Concepts*: SAS System Concepts: SAS Variables: SAS Variable Lists.

SQL has a special keyword, DISTINCT, to specify that duplicate rows are to be eliminated. The keyword appears in the SELECT statement or clause, immediately following SELECT and preceding the list of columns. So the SQL code to eliminate duplicates from our table is:

```
PROC SQL;
CREATE TABLE sex_age_distinct AS
SELECT     DISTINCT *
FROM       sex_age
;
QUIT;
```

It produces the same output as PROC SORT (Exhibit 2-12).

Preview: We will later (in Section 3.2) see a different but somewhat similar usage of DISTINCT to modify the vertical evaluation of summary statistics.

Note the absence of an ORDER BY clause, which means that the SQL processor is not obligated to return the rows in any particular order. In this situation, the processor will store the rows in the order they happen to be in after completion of necessary processing. The process of identifying distinct records entails sorting, and because the asterisk was used in the SELECT statement to include all columns, the precedence of the columns as sort keys followed their internal order, just as it did when the special SAS name list _ALL_ was used in the BY statement of the PROC SORT step. Consequently, we not only get the same rows, but we also get them in the same order. However, the ordering in the PROC SORT solution was essential in that it was a prerequisite to use of the NODUPRECS option, whereas in the SQL solution it is more of a side effect.

2.11 Summary

We have seen that SELECT can either stand alone as a statement, in effect serving as a report generator, or serve as a clause within a CREATE statement. It can operate on some or all columns from an existing source (designated using the mandatory FROM clause), and it can derive new columns. These derivations can be in the form of totals and other summary statistics. Derivations can also be conditional. Other basic capabilities include filtering, elimination of duplicate rows, and reordering of rows prior to presentation or storage.

The following clauses are subordinate to the SELECT statement or clause: FROM, WHERE, GROUP BY, HAVING, and ORDER BY. Only FROM is required, but if others are used they must appear in the order stated.

Preview: Later we'll see some other clauses that appear within a SELECT clause or statement. The ON clause (see Section 4.3) is subordinate to the FROM clause, and so appears after FROM and before WHERE. The INTO clause (see Section 8.2) precedes the FROM clause. The USING clause (see Section 10.5) is found only in CREATE VIEW statements and is the only clause that follows ORDER BY.

We looked at non-SQL counterparts to most of the SQL statements presented in this chapter. Each of those counterparts consisted of a single DATA or PROC step. In the next chapter, we'll look at SQL features that are not inherently much more complicated than the ones we've already seen, but whose non-SQL counterparts involve multiple steps.

C h a p t e r **3**

More Building Blocks

The previous chapter presented what we called "basic building blocks." Each of them was illustrated with a bit of SQL code and compared with equivalent or near-equivalent non-SQL SAS code consisting of one and only one step.

In this chapter, we add a few more blocks. The main difference is that here the non-SQL SAS counterparts involve more than one step.

Let's begin by creating a table we will use in the examples:

```
PROC SQL;
CREATE TABLE teens AS
SELECT      name AS FName,
            age
FROM        sashelp.class
WHERE       age>12
;
QUIT;
```

The table looks like Exhibit 3-1.

Exhibit 3-1 TEENS

FName	Age
Alfred	14
Alice	13
Barbara	13
Carol	14
Henry	14
Janet	15
Jeffrey	13
Judy	14
Mary	15
Philip	16
Ronald	15
William	15

3.1 Combining Summary Statistics with Original Detail

We learned in the previous chapter (see Section 2.6) how to compute summary statistics. So, for example, we could compute, from TEENS, how large each age cohort is (that is, how many rows have AGE=13, how many have AGE=14, and so on). But such a summary table has one row for each value of AGE; in other words, it contains **only** the summary data. What if we need a row for **each** NAME, containing a combination of the data in the TEENS table and the cohort sizes? For example, we want to see a row for Barbara showing her name, her age (13), and the total number of 13-year-olds (3).

Avoiding (for now) SQL, we could begin by calling on PROC FREQ to get the cohort sizes. The code is:

```
PROC FREQ DATA=teens NOPRINT;
TABLES age / OUT=cohorts(DROP=percent RENAME=(count=Many) );
RUN;
```

The output data set (COHORTS) is shown in Exhibit 3-2.

Exhibit 3-2 COHORTS

Age	Many
13	3
14	4
15	4
16	1

Reference: Read more about the FREQ Procedure in the *Base SAS 9.2 Procedures Guide*: Procedures: The FREQ Procedure or in Section 4.11 of *The Little SAS Book* (Fourth Edition).

To combine these counts with the original data, we first sort that original data:

```
PROC SORT DATA=teens OUT=sorted;
BY age;
RUN;
```

Exhibit 3-3 presents the result.

Exhibit 3-3 SORTED

FName	Age
Alice	13
Barbara	13
Jeffrey	13
Alfred	14
Carol	14
Henry	14
Judy	14
Janet	15
Mary	15
Ronald	15
William	15
Philip	16

Then we combine the original data with the counts, via a MERGE statement:

```
DATA detail_and_counts;
MERGE sorted cohorts;
BY age;
RUN;
```

We now have all of the data together, but the names are grouped by AGE and thus not in alphabetical order, as we see in Exhibit 3-4.

Exhibit 3-4 DETAIL_AND_COUNTS (before sorting)

FName	Age	Many
Alice	13	3
Barbara	13	3
Jeffrey	13	3
Alfred	14	4
Carol	14	4
Henry	14	4
Judy	14	4
Janet	15	4
Mary	15	4
Ronald	15	4
William	15	4
Philip	16	1

So we sort again to restore the original alphabetical order:

```
PROC SORT DATA=detail_and_counts;
BY fname;
run;
```

Exhibit 3-5 reflects the result.

Exhibit 3-5 DETAIL_AND_COUNTS (after sorting)

FName	Age	Many
Alfred	14	4
Alice	13	3
Barbara	13	3
Carol	14	4
Henry	14	4
Janet	15	4
Jeffrey	13	3
Judy	14	4
Mary	15	4
Philip	16	1
Ronald	15	4
William	15	4

It has taken four steps to get this output. In contrast, using SQL, we can simply write:

```
PROC SQL;
CREATE TABLE detail_and_counts AS
SELECT      fname,
            age,
            COUNT(*) AS Many
FROM        teens
GROUP BY    age
ORDER BY    fname
;
QUIT;
```

This produces the same result (seen in Exhibit 3-5) as the non-SQL SAS code.

Tip: Using the asterisk (*) as the argument of the COUNT function causes the SQL processor to count the number of rows in the group. With any other argument, the function would count the number of non-missing values of that argument.

This SQL statement uses only language features that were demonstrated in the previous chapter. However, note that the SELECT list includes both original variables (FNAME and AGE) **and** a summary statistic (COUNT). Most implementations of SQL do not permit this, but PROC SQL does. The feature is called *remerging*, and the log generated by this CREATE statement, or by any statement that uses this feature, says:

```
NOTE: The query requires remerging summary statistics
back with the original data.
```

The remerging feature is called into use implicitly, simply by including an item in the SELECT list that comes from the original detail present in the source data.

Tip: SAS 9.2 introduces new options, REMERGE and NOREMERGE, which can be used to control the availability of the remerging feature. The options are available at both the SAS system level and the SQL procedure level.

3.2 Summary Statistics Based on Distinct Values

Sometimes, when we need summary statistics derived from our data, we want the computer to ignore repetition of values. For example, suppose we want to know the average of the AGE values that occur in our TEENS table, ignoring repetitions of those values. In other words, we need an unweighted mean of AGE, in the sense that we want to include each particular value (13 and so on) only once, no matter how many times it may appear.

The weighted mean is pretty simple to calculate, with or without SQL. The non-SQL code is:

```
PROC MEANS DATA=teens MEAN MAXDEC=3;
VAR age;
RUN;
```

and the result is:

```
The MEANS Procedure

Analysis Variable : Age

       Mean

------------
     14.250
------------
```

Deriving our unweighted mean via PROC MEANS is more complicated, and is a two-step proposition. First we have to eliminate repetitions of AGE values; one way to do this is with PROC FREQ:

```
PROC FREQ DATA=teens NOPRINT;
TABLES age / out=freq2means(KEEP = age);
RUN;
```

The output of this step is the intermediate data set, FREQ2MEANS, shown in Exhibit 3-6.

Exhibit 3-6 FREQ2MEANS

Age
13
14
15
16

Now we can proceed to find the average of these distinct (unduplicated) AGE values, using PROC MEANS:

```
PROC MEANS DATA=freq2means MEAN MAXDEC=3;
VAR age;
RUN;
```

and get the following output:

```
The MEANS Procedure

Analysis Variable : Age

      Mean
------------
     14.500
------------
```

This derivation can be done in just one PROC SQL statement. We can even display the simple weighted mean alongside. The code is:

```
PROC SQL;
SELECT          MEAN(          age)
                 LABEL =    'Weighted' FORMAT=8.3,
                MEAN(DISTINCT age)
                 LABEL = 'Unweighted' FORMAT=8.3
FROM            teens
;
QUIT;
```

and it generates this result:

```
Weighted  Unweighted
------------------
 14.250     14.500
```

The use of the keyword DISTINCT here opens the possibility for some confusion, because we saw another usage for the same keyword in the previous chapter, in the context SELECT DISTINCT (see Section 2.10). Recall that the earlier usage brought about the elimination of duplicate **rows**, whereas the usage we have just introduced eliminates duplicate values from entering into the computation of summary statistics.

Let's see what happens if we attempt to exercise the row-oriented usage of DISTINCT to solve our present problem. The code would be:

```
PROC SQL;
SELECT          DISTINCT MEAN (age) LABEL='DISTINCT MEAN (age)'
FROM            teens
;
QUIT;
```

and the output would be:

```
DISTINCT
  MEAN
  (age)
--------
  14.25
```

To produce this result, the SQL processor first considers the vector of AGE values that it gets from the source (the table TEENS). To see this vector, we can run the simple query:

```
PROC SQL;
SELECT      age
FROM        teens
 ;
QUIT;
```

which gives us:

```
    Age
--------
     14
     13
     13
     14
     14
     15
     13
     14
     15
     16
     15
     15
```

There are 12 values of AGE here, from the 12 rows in the source. So the MEAN function returns a single row containing the weighted mean (14.25), and only then does the DISTINCT keyword go to work to eliminate duplicates. In this situation, however, there are no duplicates to eliminate, and the result is the (incorrect) value 14.25.

Instead, we coded the keyword DISTINCT immediately before the function argument. This caused the SQL processor to eliminate duplicates **before** averaging, rather than after. Thus, the result of the mean computation was 14.5 (correct) rather than 14.25 (incorrect).

Of course, both usages of DISTINCT are useful. They just do different things, so it's important to pick the appropriate one for a particular task.

3.3 Preprocessing the Source with Inline Views

So far in this chapter we've seen two PROC SQL features (remerging of summary statistics with detail and the DISTINCT restriction for summary statistics) that allow a single PROC SQL statement to accomplish what might take two or more non-SQL steps.

We now conclude the chapter by looking at the inline view, a tool that allows what would otherwise be a multi-statement SQL process to be telescoped into a single SQL statement.

Preview: An inline view is a particular type of view. The subject of named views, and views in general, is taken up in Chapter 10.

To illustrate the development and use of inline views, let's continue with our example. Suppose we now want to identify the **largest** cohort(s) in our population of teens. In other words, at which age level (or levels, since ties are a possibility) are there the most individuals?

Once again we can develop a non-SQL SAS solution by starting with our PROC FREQ code to list the cohorts and their sizes. This time we invoke the ORDER=FREQ option so that the cohorts will appear from largest to smallest (thus eliminating the need for a subsequent PROC SORT step):

```
PROC FREQ DATA=teens NOPRINT ORDER=FREQ;
TABLES age / OUT=highlow(DROP=percent RENAME=(count=Many) );
RUN;
```

This yields the HIGHLOW table (shown in Exhibit 3-7).

Exhibit 3-7 HIGHLOW

Age	Many
14	4
15	4
13	3
16	1

Now we use a DATA step to move through the observations until a decline is detected; doing this ensures that ties will be included in the result. The code is:

```
DATA largest;
SET highlow;
IF many < LAG(many) THEN STOP;
RUN;
```

The resulting data set looks like Exhibit 3-8.

Exhibit 3-8 LARGEST

Age	Many
14	4
15	4

Tip: In solving problems such as this, SQL tends to naturally admit ties, whereas admitting ties tends to take a bit of extra effort with non-SQL SAS techniques. On the other hand, when you want to ignore ties and select single "winners," it's non-SQL tools that usually have the advantage.

So we've determined that the largest cohorts are the 14-year-olds and the 15-year-olds, with 4 individuals in each group. It's taken two steps to derive this result.

With techniques we've already seen, we can do the same thing in two SQL statements. The first statement is:

```
PROC SQL;
CREATE TABLE temp AS
SELECT     age,
           count(*) AS Many
FROM       teens
GROUP BY   age
  ;
```

This gets the cohort sizes. Exhibit 3-9 reflects the table TEMP.

Exhibit 3-9 TEMP

Age	Many
13	3
14	4
15	4
16	1

Second, we extract the largest cohorts by running this code:

```
CREATE TABLE largest AS
SELECT       *
FROM         temp
HAVING       many = MAX(many)
;
QUIT;
```

The MAX function returns the value 4 (corresponding to both AGE=14 and AGE=15). The HAVING clause then performs comparisons with all the rows and finds the equality condition to be true for (of course) the rows where AGE=14 or AGE=15. So the result, seen in Exhibit 3-10, is identical to what we got when we avoided SQL.

Exhibit 3-10 LARGEST

Age	Many
14	4
15	4

The reason we needed two statements is that we needed one summary statistic (COUNT) **with** grouping and another (MAX) **without** grouping. Any time the grouping requirements for summary statistics are not the same, it's impossible to simply call for them in the same SELECT statement or clause.

Fortunately, SQL offers a way to integrate these two statements into one and thus avoid parking intermediate results in a table. The device is called an *inline view*, and it is implemented by having the source for one SELECT be a pair of parentheses containing another SELECT (excluding any ORDER BY clause as well as the closing semicolon).

So we can take our second statement, which was:

```
CREATE TABLE largest AS
SELECT       *
FROM         temp
HAVING       many = MAX(many)
;
```

replace the name of the source table with a pair of parentheses:

```
CREATE TABLE largest AS
SELECT       *
FROM         ()
HAVING       many = MAX(many)
;
```

then insert our first query:

```
SELECT       age,
             count(*) AS Many
FROM         teens
GROUP BY     age
```

into the parentheses, giving us this to run:

```
PROC SQL;
CREATE TABLE largest AS
SELECT       *
FROM         ( SELECT        age,
                             count(*) AS Many
                 FROM        teens
                 GROUP BY    age
               )
HAVING       many = MAX(many)
;
QUIT;
```

The results are the same as what we got from the two-statement solution; see, again, Exhibit 3-9. In essence, the inline view is a preprocessor that reads the ultimate source (table TEENS in this example) and performs SQL processing of the data before handing it up to the outer query.

Tip: Inline views can be nested, so that the source for an inline view is another inline view.

3.4 Summary

In this chapter, we have added three tools to our collection. One is the remerging feature, which automatically integrates aggregate data (produced by summary statistics) with original detail. The second is the DISTINCT keyword, which is employed to eliminate duplicate data before it is processed by a summary function. The third is the inline view, which is a flexible tool for performing in one statement SQL processing that would otherwise require multiple statements.

These three tools differ from those seen earlier in that their non-SQL SAS equivalents require multiple steps. However, all of the techniques we've seen in this chapter and the chapter preceding it have something in common: they process data from a **single** source. Even the inline view usage we have seen has this characteristic, because it has just one ultimate source.

In the next chapter and the two that follow it, we'll look at SQL techniques for integrating data from **multiple** sources.

C h a p t e r 4

Joins

Each of the SQL statements we have examined so far has drawn data from a single source. However, much of the power of SQL comes from its capacity to combine data from two or more sources. In this chapter and the two that follow, we look at the SQL features that integrate data. We begin in this chapter with joins, which are probably the most heavily used SQL device for data integration.

Roughly speaking, the join is the SQL counterpart to the DATA step MERGE statement. They both combine data from two (or more) tables or views, in a horizontal (side-by-side) fashion. There are also many differences, and we devote most of this chapter to comparisons and contrasts.

4.1 Avoiding Ambiguity in Column References

Combining data from multiple tables opens up the possibility of duplicate column names. That has not been a problem up until now. We've been able to process statements like:

```
PROC SQL;
SELECT      name
FROM        sashelp.class
WHERE       name='Jane'
;
QUIT;
```

without concern about NAME being ambiguous, because no two columns in a table can have the same name. However, we can be more precise by coding:

```
PROC SQL;
SELECT      class.name
FROM        sashelp.class
WHERE       class.name='Jane'
;
QUIT;
```

which makes it explicit that we want the column NAME from the table CLASS. We don't need that precision in this context; we're just demonstrating its availability.

To be as explicit as possible, we might attempt to code:

```
PROC SQL;
SELECT      sashelp.class.name
FROM        sashelp.class
WHERE       sashelp.class.name='Jane'
;
QUIT;
```

However, this three-part notation constitutes a syntax error and is not accepted. Instead, we can use what is called an *alias* for the table, introduced by the keyword AS, as in:

```
PROC SQL;
SELECT      s_h_c.name
FROM        sashelp.class AS s_h_c
WHERE       s_h_c.name='Jane'
;
QUIT;
```

Again, this is not something that's needed here, but the technique would be needed for a join of two tables with the same name from two different libraries.

Preview: Using aliases also becomes necessary when joining a table with itself. We'll see later (in Section 13.2) that this is a very useful technique.

While we're talking about techniques for referencing columns, we might as well deal with problems that can arise when a new column is created by including a formula in the SELECT statement.

Consider this code:

```
PROC SQL;
SELECT          LOWCASE(name) AS name
FROM            sashelp.class
WHERE           name='jane';
;
QUIT;
```

We get this note in the log:

```
NOTE: No rows were selected.
```

We want the WHERE clause to examine our new NAME column, in which the uppercase letters have been converted to lowercase. Instead, it is looking at the NAME column found in the source table, which of course still has capitalization. Perhaps we can avoid this difficulty by giving the new column a distinct name, as in:

```
PROC SQL;
SELECT          LOWCASE(name) AS lowname
FROM            sashelp.class
WHERE           lowname='jane';
;
QUIT;
```

The result is:

```
ERROR: The following columns were not found in the
contributing tables: lowname.
```

This happens because of the order in which the SQL processor performs the internal steps needed to derive the results of the SELECT statement. Such details are pretty much hidden from us, and ordinarily are not matters of concern. However, in this case we have to understand that evaluation of formulas in the SELECT list is one of the last operations. A consequence is that the name assigned to our new column is not recognized if we try to use it elsewhere in the statement. Fortunately there is a remedy: insertion of the keyword CALCULATED before any reference to a new column's name. So we can change the code to:

```
PROC SQL;
SELECT       LOWCASE(name) AS name
FROM         sashelp.class
WHERE        CALCULATED name='jane';
;
QUIT;
```

When we run this code we get the expected result:

```
Name
--------
jane
```

4.2 The Simplest Merges and Joins

We begin our exploration of joins by setting up some very simple tables and using them to demonstrate the simplest merges and joins. This will help us to see the essential nature of joins and the essential difference between merges and joins.

To create the test tables we run:

```
DATA one;
DO Value1 = 11,12;
   OUTPUT;
   END;
RUN;
DATA two;
DO Value2 = 21,22,23;
   OUTPUT;
   END;
RUN;
```

This produces the tables ONE (Exhibit 4-1):

Exhibit 4-1 ONE

Value1
11
12

and TWO (Exhibit 4-2):

Exhibit 4-2 TWO

Value2
21
22
23

We can merge these with this code:

```
DATA combined;
MERGE one two;
RUN;
```

This gives us the table shown in Exhibit 4-3.

Exhibit 4-3 COMBINED (from DATA step)

Value1	Value2
11	21
12	22
.	23

So we can see the essence of what the MERGE statement does: It combines data sets in a side-by-side fashion and attempts to pair observations. Missing values arise when one of the tables has fewer rows than the other. Here the result has three rows, that being the larger of the populations of the source tables.

Reference: Read more about the MERGE statement in the *SAS 9.2 Language Reference: Dictionary* and Sections 6.4 through 6.6 of *The Little SAS Book* (Fourth Edition).

In practice, the MERGE statement is usually used in conjunction with a controlling BY statement. Usage without a BY statement, as in this example, is not common. We have exercised it solely to draw a contrast between the MERGE statement and PROC SQL joins.

So we proceed to join the same two tables:

```
PROC SQL;
CREATE TABLE combined AS
SELECT      *
FROM        one CROSS JOIN two
;
QUIT;
```

The keyword CROSS indicates that no conditions are being stipulated to control the join. It thus constitutes the simplest type of join. The result is shown in Exhibit 4-4.

Exhibit 4-4 COMBINED (from PROC SQL)

Value1	Value2
11	21
11	22
11	23
12	21
12	22
12	23

We have again combined the two tables horizontally, but this time the rows have been crossed to form what is called a Cartesian product. Each row in the first table is combined with each row in the second. The result has six rows, that being the product of the populations of the source tables.

So, both joins and merges combine data in a side-by-side fashion, typically producing rows that are longer than those in the underlying tables. The fundamental difference is found in the way individual rows are linked. Merges pair rows, so that one row from one underlying table is combined with **one** row from the other underlying table. Joins cross rows, so that one row from one underlying table is combined with **each** row from the other underlying table. We will see that the two languages (SQL and the DATA step language) have features that let us control the behavior of merges and joins. Application of such features, together with properties of the tables being joined, often makes it possible for joins and merges to emulate each other despite their fundamental differences.

4.3 Matching with Nonrepeating Keys

In the examples we've seen thus far in this chapter, the linking of the rows from the two tables has been performed according to rules built into the software (pairing for the DATA step merge, Cartesian crossing for the SQL join). The linking did not in any way involve inspecting the values of any of the variables. Such processing is useful for some purposes, but more often what one needs is *key* columns; that is, columns whose values will be matched, with the results of the comparisons controlling the linkage.

We'll illustrate with an example, but first we must build two more tables:

```
DATA u1;
INPUT Key $ Value1;
CARDS;
A 11
B 12
;
DATA u2;
INPUT Key $ Value2;
CARDS;
C 23
A 21
;
```

The column named KEY, common to both tables, is our key. Notice that within each table, there is no repetition of KEY values; each value (A or B or C) appears at most once in each table. For much of this chapter we hold to this special case, as it simplifies things a great deal, yet allows us to explore a lot of the behavior of joins. Later, we relax this restriction and take a more general look at things.

Since DATA step merges require that data sets be sorted or indexed, we take care of that right away:

```
PROC SORT DATA=u1 OUT=sorted1;
BY key;
RUN;
PROC SORT DATA=u2 OUT=sorted2;
BY key;
RUN;
```

We now have the tables SORTED1 (Exhibit 4-5):

Exhibit 4-5 SORTED1

Key	Value1
A	11
B	12

and SORTED2 (Exhibit 4-6):

Exhibit 4-6 SORTED2

Key	Value2
A	21
C	23

Symmetrically Inclusive Joins

We start with the simplest match merge (that is, a DATA step with a MERGE statement controlled by a BY statement). The code is:

```
DATA combined;
MERGE sorted1 sorted2;
BY key;
RUN;
```

The result is the data set COMBINED, which looks like Exhibit 4-7.

Exhibit 4-7 COMBINED (from DATA step)

Key	Value1	Value2
A	11	21
B	12	.
C	.	23

Because KEY=A appears in both source data sets (SORTED1 and SORTED2), the respective observations are linked and the data from the satellite (non-key) variables are brought together in a single observation. Thus the 11 and the 21 appear side-by-side. In addition, the unmatched observations are carried into the result; hence the characterization of the process as inclusive. This treatment applies to both sources (KEY=B from SORTED1 and KEY=C from SORTED2); hence the characterization of the process as symmetric. This is the default behavior of a DATA step merge.

Now let's see how we can produce the same result with SQL. To start, consider this statement:

```
PROC SQL;
SELECT      *
FROM        u1 FULL JOIN u2
ON          u1.key = u2.key
;
QUIT;
```

Notice that we're using U1 and U2 instead of the sorted counterparts (SORTED1 and SORTED2). SQL does not require presorting, so it doesn't matter.

This SELECT statement introduces two language elements we've not seen before. The ON clause (which actually should be called a sub-clause, since it is subordinate to the FROM clause) corresponds to the BY statement used with the DATA step MERGE statement; it specifies the matching condition or conditions. Notice also that we have prefixed the column names in the ON clause with table names to avoid ambiguity; this two-part notation was explained earlier (see Section 4.1).

Tip: The ON clauses in examples in this chapter call for testing equality in like-named columns. That greatly simplifies development of equivalent DATA step code (because MERGE and BY statements always bring about equality-testing of like-named variables). However, SQL is not restrictive in that way; an ON clause can contain just about any expression that evaluates to true or false.

The FULL keyword instructs the SQL processor to include, in the output, those rows in the inputs (U1 and U2 in this case) that do **not** meet the ON clause condition. Inclusion of such unmatched data is not the default in PROC SQL, as it is in the DATA step.

Our statement generates this output:

Key	Value1	Key	Value2
A	11	A	21
B	12		.
	.	C	23

We have the three rows we expect, one for the matched A rows and one each for the mismatched B and C rows. The satellite columns, VALUE1 and VALUE2, appear just as they do in the DATA step results. However, the KEY values have not been consolidated in a single column; instead, there are two of them, one from U1 and one from U2.

Recall: The DATA step's variable namespace does not permit repetition of a name, but in PROC SQL such repetition is permissible (see Section 1.1).

We can accomplish the consolidation by using the COALESCE function to create a new column, like this:

```
SELECT       COALESCE(u1.key , u2.key) AS Key,
             *
FROM         u1 FULL JOIN u2
ON           u1.key = u2.key
;
```

The COALESCE function simply evaluates its arguments, in order, and returns as its result the first one that is not a missing value. This gives us:

Key	Key	Value1	Key	Value2
A	A	11	A	21
B	B	12		.
C		.	C	23

Now we actually have three columns named KEY (one from the source table U1, one from U2, and one derived using the COALESCE function). That's not inherently a problem, and we can proceed to embed this query into a CREATE statement to store the results, like this:

```
PROC SQL;
CREATE TABLE combined AS
SELECT       COALESCE(u1.key , u2.key) AS Key,
             *
FROM         u1 FULL JOIN u2
ON           u1.key = u2.key
;
QUIT;
```

The new table (COMBINED) looks like Exhibit 4-8.

Exhibit 4-8 COMBINED (from PROC SQL)

Key	Value1	Value2
A	11	21
B	12	.
C	.	23

This is exactly the same as the output of our DATA step (see Exhibit 4-7). However, in the log we see:

```
WARNING: Variable Key already exists
         on file WORK.SQL_COMBINED.
WARNING: Variable Key already exists
         on file WORK.SQL_COMBINED.
```

There are two warning messages, one for each column named KEY after the first. The warnings arise because multiple columns with the same name can exist during SQL processing, but cannot be stored. When an attempt is made to store such like-named columns, only the first (left-most) is successfully stored; the others are rejected, as we've just seen. So our consolidated KEY column, rather than either of the other KEY columns, is stored in the new table only because we had the foresight to place it before the asterisk in the SELECT clause; that's not very dependable. Moreover, it's pretty unrefined to have those warnings in the log. So let's make the SELECT list more explicit and avoid these problems. The code becomes:

```
PROC SQL;
CREATE TABLE combined AS
SELECT       COALESCE(u1.key , u2.key) AS Key,
             value1,
             value2
FROM         u1 FULL JOIN u2
ON           u1.key = u2.key
;
QUIT;
```

The output table, which was already correct before we made this improvement to the code, is not affected.

Asymmetric Joins

A join that admits unmatched rows is known as an *outer* join. The example we've just completed admits such rows from **both** of the sources, and is for that reason called a *full* join, which is one type of outer join. In the example we have been using, A is the only value of KEY for which there is a match. B is found only in table U1, and C is found only in table U2.

What if we need only the matched rows, plus the unmatched rows from the first or left operand (source)? In our example, that would be the A and B rows. To do that in SQL, we simply qualify the word JOIN with the word LEFT, just as we earlier used FULL as the qualifier. Let's see how such a SELECT statement works with our data. The code is:

```
PROC SQL;
SELECT        *
FROM          u1 LEFT JOIN u2
ON            u1.key = u2.key
;
QUIT;
```

The result is:

```
Key          Value1  Key         Value2
----------------------------------------
A                11  A               21
B                12                   .
```

We see that we have the expected rows. Because we used the asterisk notation in the SELECT list, we get all of the columns that are encountered, including the KEY columns from both U1 and U2. As in our earlier example, we want only one KEY column when we store the results in a table. When we did a full join, we had to use the COALESCE function to consolidate the KEY columns. In this example, however, that's not necessary, because the KEY column from U1 identifies all of the rows in the result. We just have to make sure we include that KEY column and not the one from U2. So to save the results in a table, we can run:

```
PROC SQL;
CREATE TABLE combined_left AS
SELECT        u1.*,
              value2
FROM          u1 LEFT JOIN u2
ON            u1.key = u2.key
;
QUIT;
```

Here we've combined the asterisk shortcut with the prefix (qualifier) U1. We thus get all of the columns originating in U1. We complete the SELECT list by calling for VALUE2, which needs no qualifier because it occurs in U2 only. The table looks like Exhibit 4-9.

Exhibit 4-9 COMBINED_LEFT

Key	Value1	Value2
A	11	21
B	12	.

DATA step programmers know how to do the same thing with a MERGE statement. It requires adding two elements to the code we used in the counterpart to the full join: a data set option (IN=) to detect which KEY values occur in the left operand, and a subsetting IF statement. Here is the code:

```
DATA combined_left;
MERGE sorted1(IN=in1) sorted2;
BY key;
IF in1;
RUN;
```

We use the sorted versions of the inputs because the match merge has to process key values in order. The result is identical to that of the SQL left join (see Exhibit 4-9).

What if we want the "mirror-image" behavior, which would preserve all rows from the right (second-named) table but only matching rows from the left table? The adaptation of the code is pretty straightforward. In the DATA step we simply set up the filtering on the second source data set instead of the first:

```
DATA combined_right;
MERGE sorted1 sorted2(IN=in2);
BY key;
IF in2;
RUN;
```

With SQL, it is basically just a matter of changing one word; we now preface JOIN with RIGHT:

```
PROC SQL;
CREATE TABLE combined_right AS
SELECT      u2.key,
            value1,
            value2
FROM        u1 RIGHT JOIN u2
ON          u1.key = u2.key
;
QUIT;
```

We've also composed the SELECT list a bit differently. In order to follow the convention of placing the key column first, followed by the satellite columns from the first table and then the satellite columns from the second table, we've foregone the convenience of the asterisk and instead designated all of the columns explicitly.

Both the SQL right join and the DATA step code shown just before produce the same table (shown in Exhibit 4-10).

Exhibit 4-10 COMBINED_RIGHT

Key	Value1	Value2
A	11	21
C	.	23

Match MERGE versus INNER JOIN

Often it is appropriate to shed nonmatching rows from **both** sources. In SQL, this is known as an inner join. So of course we use INNER in front of JOIN, as in:

```
PROC SQL;
SELECT      *
FROM        u1 INNER JOIN u2
ON          u1.key = u2.key
;
QUIT;
```

> **Tip:** The inner join is the default join type, so we actually could omit the word "inner" and just code U1 JOIN U2.

This produces:

```
Key          Value1  Key          Value2
---------------------------------------
A              11  A             21
```

Both of the unmatched rows are excluded, so we have only the A row. As in the outer joins that we explored earlier, we have two KEY columns. They are identical, which is necessarily the case because the inner join excludes all of the rows where they differ. So when we prepare to store the results in a table and want to avoid selecting two columns with the same name, we can pick either one. One possibility is thus:

```
PROC SQL;
CREATE TABLE combined_inner AS
SELECT       u1.*,
             value2
FROM         u1 INNER JOIN u2
ON           u1.key = u2.key
;
QUIT;
```

The table (COMBINED_INNER) looks like Exhibit 4-11.

Exhibit 4-11 COMBINED_INNER

Key	Value1	Value2
A	11	21

Using the DATA step, we can accomplish the same thing by having filters on both sides of the merge, as in:

```
DATA combined_inner;
MERGE sorted1(IN=in1) sorted2(IN=in2);
BY key;
IF in1 AND in2;
RUN;
```

The result is the same as that produced using the SQL inner join and presented in Exhibit 4-11.

Summary

At this point we've finished exploring the basic varieties of the join. We've worked exclusively with tables in which key values do **not** repeat. In that special case, we've been able to develop, for each join type (full, left, right, and inner), equivalent DATA step code using the MERGE statement.

The default join type is the inner join, whereas merges by default are inclusive (that is, they behave like full joins). As a consequence, we've seen that join code tends to be more intricate for an outer join than it is for an inner join, whereas DATA steps reflect the opposite circumstance: they can be pretty streamlined when they parallel full joins and get more complicated, at least in appearance, when they are made functionally equivalent to inner joins.

4.4 Matching with Repeating Keys

The key-controlled joins we've examined so far have been restricted to tables with distinct (that is, nonrepeating) keys. We now look beyond that special case and see how joins behave when keys **can** repeat.

Obviously, we need different example tables. The following code:

```
DATA m1;
INPUT Key $ Value1;
CARDS;
A 11.1
A 11.2
B 12.1
B 12.2
;
```

produces the data set shown in Exhibit 4-12.

Exhibit 4-12 M1

Key	Value1
A	11.1
A	11.2
B	12.1
B	12.2

This code:

```
DATA m2;
INPUT Key $ Value2;
CARDS;
A 21.1
A 21.2
A 21.3
C 23.1
C 23.2
;
```

produces the data set shown in Exhibit 4-13.

Exhibit 4-13 M2

Key	Value2
A	21.1
A	21.2
A	21.3
C	23.1
C	23.2

We've constructed these tables with the KEY columns in ascending order so that we will not have to bother with sorting to prepare the data for merging.

INNER JOIN

Let's first take the DATA step code we developed to parallel the SQL inner join, and run it against our new test data:

```
DATA many_inner;
MERGE m1(IN=in1) m2(IN=in2);
BY key;
IF in1 and in2;
RUN;
```

Looking at the log, we see:

```
NOTE: MERGE statement has more than one data set with
repeats of BY values.
```

The table MANY_INNER looks like Exhibit 4-14:

Exhibit 4-14 MANY_INNER

Key	Value1	Value2
A	11.1	21.1
A	11.2	21.2
A	11.2	21.3

The data in tables M1 and M2 associated with KEY values of B and C are absent from this result because they are filtered out by the subsetting IF statement. That leaves us with only data for the KEY value A found in both M1 and M2. Because A repeats in both M1 and M2, the DATA step pairs off the rows.

We saw that same behavior at the start of this chapter (see Section 4.2), when we experimented by running a DATA step merge in the absence of a BY statement. The only thing new is that even though M1 has two A rows and M2 has three of them, there are no missing values. Instead, the third row in our result in effect inherits the value 11.2 for VALUE1 from the second row. This happens because the DATA step does not do its reinitialization processing until the beginning of a new BY group. DATA step programmers generally avoid this situation (repeating keys in more than one source data set) because it usually does not produce useful or appropriate results.

Next, we perform an inner join of these same tables:

```
PROC SQL;
SELECT       *
FROM         m1 INNER JOIN m2
ON           m1.key = m2.key
;
QUIT;
```

This yields:

```
Key          Value1  Key         Value2
---------------------------------------
A               11.1 A             21.1
A               11.2 A             21.1
A               11.1 A             21.2
A               11.2 A             21.2
A               11.1 A             21.3
A               11.2 A             21.3
```

As in earlier examples, we see the columns from the first source (here M1), followed by the columns from the second (M2). We can store the results in a table by placing the code within a **CREATE TABLE** statement, with the **SELECT** list modified to include just one of the two identical **KEY** columns, like this:

```
PROC SQL;
CREATE TABLE many_inner AS
SELECT       m1.*,
             value2
FROM         m1 INNER JOIN m2
ON           m1.key = m2.key
;
QUIT;
```

We get the table we see in Exhibit 4-15.

Exhibit 4-15 MANY_INNER

Key	Value1	Value2
A	11.1	21.1
A	11.2	21.1
A	11.1	21.2
A	11.2	21.2
A	11.1	21.3
A	11.2	21.3

Because it's an inner join, there are no rows corresponding to nonmatching KEY values (B and C). The rows that do match (that is, the A rows), have been crossed in Cartesian fashion. Each VALUE1 (11.1, 11.2) is in a row with each VALUE2 (21.1, 21.2, 21.3). This too is reminiscent of what we saw at the beginning of the chapter (see Section 4.2) when we demonstrated the cross join.

So, when **both** tables have repeating keys, joins and merges produce fundamentally different results, and there is essentially no way to manipulate their behavior to eliminate the differences. This is territory where SQL and the DATA step are simply different.

However, if repetition of keys occurs in just **one** of the source tables, the SQL inner join and the DATA step merge can be made to work equivalently. We can demonstrate by merging U1 (in which keys do not repeat; see Exhibit 4-16)

Exhibit 4-16 U1

Key	Value1
A	11
B	12

with M2 (which has repeating keys; see Exhibit 4-13). The code is:

```
DATA one_many_inner;
MERGE u1(IN=in1) m2(IN=in2);
BY key;
IF in1 AND in2;
RUN;
```

and it gives us the table seen in Exhibit 4-17.

Exhibit 4-17 ONE_MANY_INNER

Key	Value1	Value2
A	11	21.1
A	11	21.2
A	11	21.3

The subsetting IF statement excludes the nonmatching rows—that is, the B and C rows. The one A row in U1 is paired with the first A row in M2, producing the first row in the output. The other A rows in M2 are not paired, but they are included in the output and inherit the 11 in the VALUE1 column.

Now we try doing an inner join with the same data. We run:

```
PROC SQL;
CREATE TABLE one_many_inner AS
SELECT       u1.*,
             value2
FROM         u1 INNER JOIN m2
ON           u1.key = m2.key
;
QUIT;
```

The result is the same (see, again, Exhibit 4-17), but the process is different. The SQL processor, like the DATA step, eliminates the nonmatching B and C rows. Then it crosses the one A row in U1 with the three A rows in M2. Thus each row in the result is built with an 11 in the VALUE1 column; there is no inheritance going on (there can't be, because SQL treats rows as an unordered set).

OUTER JOIN

There is not much to say about outer (FULL, LEFT, or RIGHT) joins when there are repeating keys. Such joins work as they do when keys are distinct, in that unmatched rows from either or both of the source tables (depending on the type of outer join) are carried into the result. Let's look at a one-to-many full join. We take our last example, replace INNER with FULL, and adapt the SELECT list to use the COALESCE function (which we've found to be necessary for full joins). Our code is:

```
PROC SQL;
CREATE TABLE one_many_outer AS
SELECT      coalesce(u1.key, m2.key) AS Key,
            value1,
            value2
FROM        u1 FULL JOIN m2
ON          u1.key = m2.key
;
QUIT;
```

The result is shown in Exhibit 4-18.

Exhibit 4-18 ONE_MANY_OUTER (from PROC SQL)

Key	Value1	Value2
A	11	21.3
A	11	21.1
A	11	21.2
B	12	.
C	.	23.2
C	.	23.1

The A rows are the same as those that appear in the inner join. In addition, we have the unmatched rows from both U1 and M2, with nulls (missing values) arising where the source tables do not provide values. All in all, the content of this table is for the most part predictable. Notice that the rows are in an order that reflects no obvious rule. That is a consequence of the internal operation of the SQL processor in evaluating the query; in the absence of an ORDER BY clause, the processor is not obligated to deliver results in any particular order.

We can devise an analogous DATA step. It is:

```
DATA one_many_outer;
MERGE u1 m2;
BY key;
RUN;
```

It yields the data set that appears in Exhibit 4-19.

Exhibit 4-19 ONE_MANY_OUTER (from DATA step)

Key	Value1	Value2
A	11	21.1
A	11	21.2
A	11	21.3
B	12	.
C	.	23.1
C	.	23.2

This differs from the SQL result only in the ordering of the rows.

4.5 More about Joins and Merges

The examples presented thus far in this chapter have been simplified in order to keep the focus on the essential nature of joins and merges. Before concluding, we ought to generalize a bit.

Three or More Sources

We have shown only cases where two data sets (tables) are combined. However, both joins and merges can operate on three or more inputs. Here's a DATA step example:

```
DATA from3;
MERGE sorted1 m1(RENAME=(value1=Tenths) ) sorted2;
BY key;
RUN;
```

The output looks like Exhibit 4-20.

Exhibit 4-20 FROM3

Key	Value1	Tenths	Value2
A	11	11.1	21
A	11	11.2	21
B	12	12.1	.
B	12	12.2	.
C	.	.	23

Similarly, we can run this SQL code:

```
PROC SQL;
CREATE TABLE sql_from3 AS
SELECT u1.*,
       m1.value1 as Tenths,
       u2.value2
FROM   (u1 JOIN m1 ON u1.key=m1.key)
          JOIN u2 ON u1.key=u2.key
;
QUIT;
```

It gives us the table in Exhibit 4-21.

Exhibit 4-21 SQL_FROM3

Key	Value1	Tenths	Value2
A	11	11.1	21
A	11	11.2	21

The two results differ because we've let the DATA step default to a merge that includes unmatched observations, while we've let SQL default to an inner join (note the absence of any join type specification before the word JOIN). Using techniques we've seen earlier in this chapter, we could have made either program equivalent to the other.

Composite Keys

All of the matching we've done has been based on simple (single-column) keys. However, both the DATA step and SQL support composite (multicolumn) keys. To illustrate, suppose we were combining data from two tables, each of which covered many companies and many years. In the DATA step, we would simply name both of these variables in the BY statement, as in:

```
BY company year;
```

assuming of course that both data sets were appropriately sorted. In doing an SQL join, we would implement the composite key by imposing comparisons on both of the columns, linked with an AND, giving us:

```
ON          table1.company = table2.company AND
            table1.year     = table2.year
```

where TABLE1 and TABLE2 are the names or aliases of the tables being joined.

4.6 More about Joins

We have now completed the side-by-side comparison of joins and merges, which is the main purpose of this chapter. There are, however, a few more things to be said about joins alone.

There are a couple of shortcuts, natural joins and implicit joins, that can be used. They don't let us do anything we could not do without them, but they streamline the coding somewhat.

Natural Joins

The so-called *natural join* is a device that allows a lot of the specifications for a join to be established by default, thus reducing the amount of explicit coding required. Basically, in a natural join, all like-named columns are assumed to be keys to be matched. The natural join automatically coalesces the values from these key columns and then discards the original columns. No ON clause is needed.

Tip: To employ natural joins successfully, make sure that like-named columns are compatible in terms of data type (character or numeric) and make sure that satellite columns (those that are not to be matched) have distinctive names.

To illustrate, consider this statement, which appeared earlier in this chapter:

```
PROC SQL;
SELECT        COALESCE(u1.key , u2.key) AS Key,
              value1,
              value2
FROM          u1 FULL JOIN u2
ON            u1.key = u2.key
;
QUIT;
```

This is an equivalent statement:

```
PROC SQL;
SELECT        *
FROM          u2 NATURAL FULL JOIN u1
;
QUIT;
```

Implicit Joins

For inner joins only, there is an alternative syntax that actually omits the word "join."
Instead, it uses commas to imply the join action. It also omits the ON clause. The
matching conditions still have to be declared, but they are placed in the WHERE clause
(linked via AND with any other WHERE conditions that might be required).

To illustrate, consider the three-table join we presented a bit earlier:

```
PROC SQL;
SELECT u1.*,
       m1.value1 as Tenths,
       u2.value2
FROM   (u1 JOIN m1 ON u1.key=m1.key)
           JOIN u2 ON u1.key=u2.key
;
QUIT;
```

It is equivalent to:

```
PROC SQL;
SELECT u1.*,
       m1.value1 as Tenths,
       u2.value2
FROM   u1,  m1,  u2
WHERE  u1.key=m1.key AND u1.key=u2.key
;
QUIT;
```

The implicit version is a bit more streamlined, though the advantage is less obvious when there are just two input tables.

Generality of Join Conditions

All of the joins we have demonstrated in this chapter (with the exception of the simple cross join) have involved equality comparisons made on like-named keys. That is a common arrangement, and it has advantages (as we have just seen in our look at the natural join). The main reason we concentrated on that type of join was to enable us to construct equivalent DATA step code; DATA step merges work exclusively with equality matching, and BY statements do not accommodate naming differences. SQL is far more flexible. Conditions declared in the ON clause are not limited to equality comparisons and do not require corresponding, like-named columns. Any condition that can be evaluated as true or false is suitable. For example, this is a perfectly sound ON clause:

```
ON UPCASE(t1.aa) GT SUBSTR(t2.bb) OR t2.cc='##';
```

Tip: Joins based on equality conditions (called equijoins) tend to be more efficient. That is, they often can be evaluated much more quickly than other joins. See Section 13.2 for an example.

4.7 Summary

The SQL join, like the DATA step merge, is a tool for horizontal integration of data from two or more sources. Both tools are usually (but not necessarily) used with key values to specify matching requirements. When keys are not unique, joins and merges tend to diverge in their behavior.

C h a p t e r 5

Subqueries

In the last chapter, we explored at some length the ways that joins can combine information from two (or more) sources. In this chapter we look at another device, the subquery, which can also introduce information from an additional source.

We saw that in a join, the multiple source tables are introduced on a par. In contrast, a subquery is, as the name suggests, a subordinate entity. This distinction is a bit abstract, but the examples in this chapter should make it clear.

Tip: Joins and subqueries are not mutually exclusive. A query that uses a join can also include subqueries. A subquery can even invoke a join.

What does a subquery look like? It is simply a SELECT clause with subordinate clauses (FROM, WHERE, GROUP BY, and HAVING), all contained within a pair of parentheses. So it might look like this:

```
( SELECT name FROM sashelp.class WHERE age>12 )
```

This code should seem familiar, because syntactically it is exactly what we earlier called an *inline view* (see Section 3.3). An inline view, however, replaces a table reference in a FROM clause. Thus it does not introduce an additional source table; instead, it makes a source indirect rather than direct. So if this construct (a SELECT clause within a pair of parentheses) occurs as an operand in a FROM clause, it is an inline view; in other contexts, it is a subquery. Another point of contrast is that a subquery can return only one column.

As usual, we explore subqueries through examples and comparisons with DATA step techniques.

To prepare data for these examples, we generate two tables. The first is a subset of SASHELP.CLASS. The code is:

```
PROC SQL;
CREATE TABLE classgirls AS
SELECT       *
FROM         sashelp.class(RENAME=(name=FName) )
WHERE        sex='F'
;
QUIT;
```

The table CLASSGIRLS looks like Exhibit 5-1.

Exhibit 5-1 CLASSGIRLS

FName	Sex	Age	Height	Weight
Alice	F	13	56.5	84.0
Barbara	F	13	65.3	98.0
Carol	F	14	62.8	102.5
Jane	F	12	59.8	84.5
Janet	F	15	62.5	112.5
Joyce	F	11	51.3	50.5
Judy	F	14	64.3	90.0
Louise	F	12	56.3	77.0
Mary	F	15	66.5	112.0

Our second table is created by running:

```
DATA moregirls;
INPUT FName $ Age;
CARDS;
Susan    16
Jane     12
Abigail  13
Zelda    16
;
```

Exhibit 5-2 displays the result.

Exhibit 5-2 MOREGIRLS

FName	Age
Susan	16
Jane	12
Abigail	13
Zelda	16

5.1 Contexts That Expect Subqueries

SQL includes some devices that are specifically designed to work with subqueries: the EXISTS condition, and the ANY and ALL keywords (which are used in conjunction with comparison operators). When those devices are used, SQL expects to see a subquery. In contrast, in numerous contexts subqueries are allowed, but not required; we get to that situation later in this chapter. Finally, there is the IN condition, a hybrid that is usually used with a subquery but can also be used with a hardcoded list.

Let's specify a task that we can perform with the DATA step or with PROC SQL, using in turn each of the four devices: EXISTS, IN, ANY, and ALL. The task is to apply a WHERE filter that will pass along the rows for those girls in the MOREGIRLS table who are already in the CLASSGIRLS table. A quick glance at the tables tells us that the subset in question includes Jane and only Jane, but we will (repeatedly) make SAS discover that. The DATA step code should follow this form:

```
DATA already;
SET moregirls;
WHERE … ;
RUN;
```

and the SQL should look like this:

```
CREATE TABLE already AS
SELECT      *
FROM        moregirls
WHERE       …
;
```

For the DATA step, the filter can be built with the PUT function and a format derived from the CLASSGIRLS data set. We want the names from CLASSGIRLS to populate the format, so we first build the data set expected by PROC FORMAT by running:

```
DATA cntl;
LENGTH label $ 1;
SET classgirls(RENAME=(fname=start)) end=last;
RETAIN label    'Y'
       fmtname 'ClassGirls'
       type    'c';
OUTPUT;
```

```
IF LAST THEN DO;
   hlo='O';
   label=' ';
   OUTPUT;
   END;
RUN;
```

> **Reference:** Read more about creating formats in the *Base SAS 9.2 Procedures Guide*:
> Procedures: The FORMAT Procedure: Example 5: Creating a Format from a Data Set.

To actually establish the format, we run:

```
PROC FORMAT LIBRARY=work CNTLIN=cntl;
RUN;
```

Finally, we insert into the DATA step a WHERE statement containing a call to the PUT function that in turn refers to our format. This restricts processing to the names in MOREGIRLS that are already in CLASSGIRLS. The code is:

```
DATA already;
SET moregirls;
WHERE PUT(fname,$classgirls.)='Y';
RUN;
```

The output is shown in Exhibit 5-3.

Exhibit 5-3 ALREADY

FName	Age
Jane	12

Indeed, Jane is the only name in common.

Notice that we had to run two SAS steps (a DATA step and a PROC FORMAT step) just to prepare, before we even turned to coding our WHERE statement. In contrast, SQL can reference the lookup table directly within the WHERE condition and without such pre-processing.

The EXISTS Condition

SQL provides a *condition* (similar to an operator), called EXISTS, which looks for a subquery as its operand and returns a value of true (1) if evaluation of that subquery returns one or more rows, or false (0) if evaluation of the subquery returns no rows. We can employ the EXISTS condition in a WHERE clause to perform the table lookup and carry out the required filtering. The code is:

```
PROC SQL;
CREATE TABLE already AS
SELECT      *
FROM        moregirls
WHERE       EXISTS
            ( SELECT *
              FROM   classgirls
              WHERE  moregirls.fname=classgirls.fname
            )
;
QUIT;
```

The result is a one-row table containing the name "Jane." See, again, Exhibit 5-3.

We can trace the logic to see how this is derived. The processor first considers "Susan," because that is the FNAME value in the first row of MOREGIRLS. Should "Susan" be in the results? That depends on the outer WHERE clause, which requires that we evaluate the subquery found within the parentheses. To do this, the processor plugs "Susan" into the subquery, making it:

```
( SELECT *
  FROM   classgirls
  WHERE  'Susan'=classgirls.fname
)
```

Because "Susan" is not found in CLASSGIRLS, no rows from CLASSGIRLS satisfy the **subquery's** WHERE clause, and the subquery returns zero rows. As a consequence, the EXISTS condition is false, so the "Susan" row of MOREGIRLS does not satisfy the **outer** WHERE condition and is therefore excluded from the results.

The process is repeated for the "Jane" row of MOREGIRLS. In that case, the subquery becomes:

```
( SELECT *
  FROM   classgirls
  WHERE  'Jane'=classgirls.fname
)
```

Because "Jane" is in one row of CLASSGIRLS, the subquery returns one row and the EXISTS condition is true. Consequently, the "Jane" row of MOREGIRLS satisfies the outer WHERE clause and becomes part (actually all, as it turns out) of the main query's results.

"Abigail" and "Zelda" are handled in the same fashion, and the results are the same as they were for "Susan." So the output table ALREADY contains just one row (for "Jane").

This is fairly complicated internally. Because the subquery refers to the outer data source (in this case via the table name MOREGIRLS), the subquery keeps changing as the individual rows of MOREGIRLS are processed. For this reason it is termed a *correlated* subquery.

Notice that we used the asterisk (*) shortcut in the subquery to select all of the columns in CLASSGIRLS. The EXISTS condition provides the only usage of subqueries in which a subquery can have more than one column. However, if we think about that a bit, we realize that it does not matter. The EXISTS condition cares only about whether or not there are any rows. It does not care about the columns. Moreover, the yield of this subquery (the rectangular array of values that it contains) is not displayed, not passed to the outer query, not stored for later use, and not used in any way. The EXISTS condition only **counts** rows, and even then it only has to count up to one. Because it is only this binary count that matters, we can place anything that is valid in the SELECT list. We could recode our statement as:

```
PROC SQL;
CREATE TABLE already AS
SELECT      *
FROM        moregirls
WHERE       EXISTS
            ( SELECT 'Hello World'
              FROM   classgirls
              WHERE  moregirls.fname=classgirls.fname
            )
;
QUIT;
```

and get the same results See, again, Exhibit 5-3. Note that unlike the typical "Hello World" demonstration, this one never displays the greeting.

The IN Condition

SQL also has a condition, IN, which (unlike EXISTS) has two operands. If the value provided by the first operand is among the (generally multiple) values provided by the second operand, the condition is deemed to be true.

The IN condition offers a solution to the problem of constructing a WHERE clause referencing CLASSGIRLS. The code is:

```
PROC SQL;
CREATE TABLE already AS
SELECT         *
FROM           moregirls
WHERE          fname IN
               ( SELECT       fname
                 FROM         classgirls
               )

;
QUIT;
```

The result, as before, is as shown in Exhibit 5-3.

Notice that the subquery does not reference any columns from the outer query. Consequently, it only needs to be evaluated once. It is thus a *noncorrelated* subquery.

Tip: To see if a subquery is noncorrelated, try running it as a separate statement. Take the code, excluding the container parentheses, and append a semicolon. Such a statement derived from a correlated subquery will fail because of the undefined reference; code from a noncorrelated subquery should work.

The behavior of the IN condition is pretty straightforward. It is satisfied if its first (left) operand (here an FNAME value from the MOREGIRLS table) matches at least one of the values in the second operand (the subquery). So, for example, when the SQL processor is considering the second row of MOREGIRLS (the one containing "Jane"), it looks at the list of names provided by the subquery, sees "Jane," and therefore keeps the row. In the cases of the other three rows of MOREGIRLS there is no such match, and the rows are excluded.

Also notice the namespace separation. The column FNAME is in both tables (MOREGIRLS and CLASSGIRLS), yet we are able to reference either column without prefixing a table name or alias. The subquery and the outer query are separate contexts. This is true of subqueries in general, and is not a consequence of the IN operator.

Unlike the EXISTS condition and the keywords ANY and ALL (discussed in the following sections), the IN condition does not require a subquery. It can also work with a list of constants. The syntax, for our example, would be:

```
PROC SQL;
CREATE TABLE already AS
SELECT         *
FROM           moregirls
```

```
WHERE           fname IN
                   ( 'Alice    ' , 'Barbara' , 'Carol   ' ,
                     'Jane     ' , 'Janet   ' , 'Joyce   ' ,
                     'Judy     ' , 'Louise  ' , 'Mary    '
                   )
   ;
QUIT;
```

However, in most real-world applications, flexibility is served by keeping lists in tables rather than hardcoding them in this fashion.

The ANY Keyword

The keyword ANY gives us yet another way of developing the WHERE clause for our query. We can run this code:

```
PROC SQL;
CREATE TABLE already AS
SELECT          *
FROM            moregirls
WHERE           fname = ANY
                   ( SELECT        fname
                     FROM          classgirls
                   )
   ;
QUIT;
```

ANY supplements a comparison operator (in this case the equality operator). The WHERE condition here is deemed to be true if it is true for at least one of the values supplied by the subquery. So processing would begin with the name "Susan" from MOREGIRLS. "Susan" is compared in turn with each name from CLASSGIRLS. In each case the equality comparison is false, so the condition (= ANY) is false and the WHERE clause therefore excludes "Susan" from the result set of the outer query. It then turns to "Jane." Because "Jane" is found in CLASSGIRLS, the equality is true for at least one row of CLASSGIRLS, making the condition true. So "Jane" is in the result set, which once again is as displayed in Exhibit 5-3.

ANY can be used with other comparison operators. To illustrate, we will for a moment put aside our task and replace the equality operator in the SELECT clause we just presented with a greater-than operator:

```
PROC SQL;
SELECT       *
FROM         moregirls
WHERE        moregirls.fname > ANY
                ( SELECT       fname
                  FROM         classgirls
                )

;
QUIT;
```

Now the process essentially amounts to considering each FNAME value from
MOREGIRLS and determining where it would fall if alphabetized with the list of names
from CLASSGIRLS. If it would appear first, then it is not greater than any of the names
from CLASSGIRLS, so the WHERE predicate would be false and the name would not be
in the result set. In our example, because the alphabetically first name in CLASSGIRLS
is Alice, that is the case only for Abigail. The other three names do result in hits from the
> ANY condition, so the result is:

```
FName            Age
------------------
Susan             16
Jane              12
Zelda             16
```

Another way of viewing this result is to observe that the rows in the result are those with
FNAME values that are higher than "Alice," the minimum (alphabetically first) FNAME
value in CLASSGIRLS.

Tip: Such comparisons can work with numeric values as well as with character values;
numeric comparisons simply involve magnitudes rather than alphabetical ordering.

The ALL Keyword

The keyword ALL, like ANY, works with a comparison operator and, as you might
guess, requires that the comparison be true for **every** row of the subquery, rather than for
just some rows. Conceivably, our example (filtering out names in MOREGIRLS that do
not match those in CLASSGIRLS) could be solved using ALL. The code would be
something like this:

```
PROC SQL;
CREATE TABLE already AS
SELECT       *
FROM         moregirls
```

```
WHERE          not( fname NE ALL
                  ( SELECT      fname
                    FROM        classgirls
                  )
               )
  ;
  QUIT;
```

The double negation makes this pretty complicated. Clearly the code employing the ANY keyword is more suitable.

However, just to see the operation of the ALL keyword, let's substitute ALL for ANY in the example with the greater-than comparison. That code becomes:

```
PROC SQL;
SELECT       *
FROM         moregirls
WHERE        moregirls.fname > ALL
               ( SELECT      fname
                 FROM        classgirls
               )
  ;
  QUIT;
```

We can again think of the evaluation as a series of trial alphabetizations. This time we are looking for names from MOREGIRLS that would appear at the end of the list of names from CLASSGIRLS. Susan and Zelda qualify (because the alphabetically highest name in CLASSGIRLS is Mary). Thus the result is:

```
FName          Age
-----------------
Susan           16
Zelda           16
```

Subqueries Outside the WHERE Clause

We've been looking at subqueries introduced by EXISTS, IN, ANY, and ALL. Such subqueries yield Boolean (true/false) results, and are most often used in WHERE clauses. That's the usage we have been demonstrating as we have created filters to apply to rows from MOREGIRLS. However, such subqueries actually can be used just about anywhere where the result makes sense. Here's an example:

```
PROC SQL;
SELECT        *
FROM          moregirls
ORDER BY      fname IN
                ( SELECT      fname
                  FROM        classgirls
                ),
              fname
  ;
QUIT;
```

This is a very simple query, except for the ORDER BY clause. The first ORDER BY item invokes the IN condition that we saw earlier. Because "Jane" is the only FNAME value common to the two tables, it returns a value of true (1) for Jane's row and false (0) for the other rows in MOREGIRLS. Because the ordering is, by default, ascending, "Jane" appears last; the other three names are alphabetized under the control of the second ORDER BY item (FNAME).

Behind the scenes, the SQL processor builds a table resembling what we would get if we ran:

```
CREATE TABLE behindscenes AS
SELECT        fname IN
                ( SELECT      fname
                  FROM        classgirls
                ),
              *
FROM          moregirls
  ;
```

It looks like Exhibit 5-4.

Exhibit 5-4 BEHINDSCENES

	FName	Age
0	Susan	16
1	Jane	12
0	Abigail	13
0	Zelda	16

This internal table is then sorted by the (unnamed) first column, with the FNAME values as the tiebreaker. Then the first column is discarded and the results are displayed, so that we see:

```
FName          Age
------------------
Abigail         13
Susan           16
Zelda           16
Jane            12
```

5.2 General Usage of Subqueries

Up until now, we've been working with subqueries introduced by one or another of the special devices EXISTS, IN, ANY, and ALL. In those contexts, it is permissible for the subquery to return any number of rows; the logic that is in effect always reduces the result to a single Boolean (true/false) value.

Subqueries can also be used, without the special devices, almost anywhere in SQL code where you can use a scalar expression. However, such employment of a subquery works correctly only if the subquery returns a scalar result (that is, a single column and no more than one row). Recall that all subqueries (except those specified in an EXISTS condition) are limited to one column, so that restriction is not new. What is new here is the restriction to one row. To see why that is necessary, consider this query:

```
PROC SQL;
SELECT    fname,
          age,
          ( SELECT fname
            FROM    classgirls
            WHERE   moregirls.age    = classgirls.age
              AND   moregirls.fname ^= classgirls.fname
          ) AS SameAge
FROM      moregirls
;
QUIT;
```

The intention is to find, for each FNAME/AGE pair in MOREGIRLS, the names of other girls in CLASSGIRLS who are the same age. When we run the code, the log shows:

```
ERROR: Subquery evaluated to more than one row.
NOTE: Correlation values are: Age=13 FName='Abigail' .
```

Looking at the listing, we see:

```
FName          Age  SameAge
--------------------------
Susan           16
Jane            12  Louise
```

There are no results for Abigail, and none for Zelda either. Let's try to see how this happened.

The first row of MOREGIRLS is for Susan, age 16. There are no 16-year-old girls in CLASSGIRLS, so the correlated subquery returns nothing and the third column of the first row of results is null. The next row in MOREGIRLS is for Jane, age 12. In this case, the subquery finds exactly one 12-year-old girl, Louise (there is a row for Jane herself in CLASSGIRLS, but the second condition in the WHERE clause filters that out). So the name "Louise" fits nicely, by itself, into the third column of the second row of output.

Next, the processor turns to Abigail, age 13. In this case, the subquery finds **two** girls the same age (Alice and Barbara). This is a problem. Although the main query is going to have many rows (one for each row of MOREGIRLS), each evaluation of the subquery is supposed to fill the third column of just a **single** row. In other words, the subquery is expected to provide a scalar value. Multiple values cannot be accommodated, hence the ERROR message.

Note that this is a run-time error condition, not detectable until processing of the data is under way. Nothing in the code can predict it, and indeed with different data in the two tables, it might not arise. Also be aware that the results that are produced are incomplete. We know this because the main source (MOREGIRLS) contains a fourth row, for Zelda; but because processing halts when the error is detected, there is no corresponding row in the results.

Clearly, this is a risky situation. It is good practice to avoid using a subquery that must return scalar results unless you can guarantee that the subquery is compliant with this restriction. There are basically two ways that you can be sure of compliance. One technique involves properties of the data. For example, an appropriate integrity constraint (a topic we consider later in Section 9.7) might prevent a subquery from returning multiple rows. That approach is not suitable here; we do not want to constrain our table to have just one girl of any particular age. The other technique focuses not on the data but rather on the query, and involves the use of grouping and summary statistics to reduce multiple rows to a single row. We can demonstrate this reduction technique by changing the code we just used to:

```
PROC SQL;
SELECT    fname,
          age,
          ( SELECT COUNT(*)
            FROM    classgirls
            WHERE   moregirls.age    = classgirls.age
              AND   moregirls.fname ^= classgirls.fname
            GROUP BY classgirls.age
          ) AS SameAge
FROM      moregirls
;
QUIT;
```

Notice the COUNT keyword and the GROUP BY clause. Now the third column is to report the **number** of girls of the same age as each girl in MOREGIRLS, and **not** their individual names. This is by definition a scalar value, not a list. That's enough to ensure that the "evaluated to more than one row" error will not occur. The results are:

```
FName          Age   SameAge
---------------------------
Susan           16        .
Jane            12        1
Abigail         13        2
Zelda           16        .
```

Nulls (missing values) arise in places where there are no rows to count. We could refine the query by employing the COALESCE function to transform these to zeroes.

5.3 Summary

A subquery is essentially a SELECT clause coded within another clause. There are two types of subqueries, correlated and noncorrelated. A correlated subquery contains one or more references to the outer query, and so must be evaluated separately for each row of the outer query. A noncorrelated subquery has no such references and does not require such re-evaluation.

Subqueries are often used with the EXISTS condition, with the IN condition, or connected to a comparison operator via either the ANY or the ALL keyword. In these contexts, subqueries might return multiple rows. Subqueries can be used elsewhere, but only if it is certain that the subquery will return a single scalar value for each row of the outer query.

Chapter **6**

Set Operators

In the two previous chapters, we explored SQL joins and subqueries at some length. These are the most widely used devices in SQL for combining data from multiple sources. However, there is another technique. In this chapter we take up what are called *set operators*.

Set operators are so designated because they are conceptually derived from mathematical set theory. There are four set operators: UNION, INTERSECT, EXCEPT, and OUTER UNION. There are also two options (ALL and CORRESPONDING) that affect the behavior of the operators.

6.1 The Contrast between Joins and Set Operators

Before we delve into PROC SQL's set operators, let's establish some basic distinctions between joins and set operators. This can be done with a simple example, starting with the creation of two tiny tables. Here's the code:

```
DATA a;
Aa = 1;
RUN;
DATA b;
Bb = 2;
RUN;
```

So each of these tables has one row and one column. We can use PROC SQL to combine the two via a simple cross join:

```
PROC SQL;
SELECT      *
FROM        a CROSS JOIN b
;
QUIT;
```

If you recall what we saw in the explanation of joins (see Section 4.2), you should be able to anticipate the result, which is:

```
    Aa          Bb
------------------
     1           2
```

Now, in contrast, we'll look at the simplest form of the most widely used of the set operators: UNION. The code to combine our two tables is:

```
PROC SQL;
( SELECT      *
  FROM        a )
UNION
( SELECT      *
  FROM        b )
;
QUIT;
```

Before we look at the effect of this statement (which is indeed a single statement, terminated by a single semicolon), let's look at the syntax and compare it to that of the join. The parentheses are not actually necessary; they are included just to emphasize the sequence in which operations are performed. Notice that UNION is inserted between two SELECT clauses (each of which has, as it must, a subordinate FROM clause). A set operator works on the **results** of **two** SELECT clauses. This is unlike a join, which is implemented **within** the FROM clause of a **single** SELECT statement. So you cannot simply substitute a UNION (or any other set operator) for a join; they operate at different levels.

Now it's time to look at the result:

```
      Aa
-------
       1
       2
```

We see the two numeric values, this time arranged vertically rather than horizontally. This reflects the fundamental difference between joins and set operators. Joins typically align rows and accrete (that is, accumulate or collect) columns; set operators align columns and (loosely speaking) accrete rows. This is something of an oversimplification, of course. SQL is not a matrix language and provides relatively little symmetry between rows and columns. So the contrast drawn here between joins and set operators is only a conceptual foundation.

6.2 Set Operators: Preview

The behavior of the four set operators depends on characteristics of the data being processed and on the options that are in effect. We go into these details at some length, but only after we look at some simple examples that convey the essence of each operator.

First, we need some test data. Once again, we can use subsets of the SASHELP.CLASS data set. If we run:

```
PROC SQL;
CREATE TABLE one AS
SELECT       name as FName, weight, age
FROM         sashelp.class
WHERE        age<13 and LENGTH(name) GE 5
ORDER BY     age, RANUNI(1)
;
CREATE TABLE two AS
SELECT       name as FName, age, height
FROM         sashelp.class
WHERE        age<13 and LENGTH(name) LE 5
ORDER BY     age, RANUNI(2)
;
QUIT;
```

we get ONE (see Exhibit 6-1):

Exhibit 6-1 ONE

FName	Weight	Age
Thomas	85.0	11
Joyce	50.5	11
James	83.0	12
Robert	128.0	12
Louise	77.0	12

and TWO (see Exhibit 6-2).

Exhibit 6-2 TWO

FName	Age	Height
Joyce	11	51.3
James	12	57.3
John	12	59.0
Jane	12	59.8

Notice that the two tables have different columns, and that common columns are not necessarily aligned in the same position (counting from left to right). Also notice the overlap: names with exactly five letters appear in both tables. Finally, notice that random numbers were used in the ORDER BY clause to shuffle the rows a bit.

Let's go ahead and exercise the four set operators. To keep things simple for now, we call for the same columns (FNAME and AGE) in the same order in our SELECT lists.

We start with the UNION operator. If we run:

```
PROC SQL;
SELECT      fname, age
FROM        one
UNION
SELECT      fname, age
FROM        two
;
QUIT;
```

we get this result:

```
FName            Age
------------------
James             12
Jane              12
John              12
Joyce             11
Louise            12
Robert            12
Thomas            11
```

The result is **inclusive** in the sense that each NAME/AGE pair that appears in **either** ONE or TWO appears in the result. However, duplicate rows (for James and Joyce) have been eliminated.

Next we turn to INTERSECT:

```
PROC SQL;
SELECT      fname, age
FROM        one
INTERSECT
SELECT      fname, age
FROM        two
;
QUIT;
```

The logic here is **exclusive**; only rows that appear in **both** operands (that is, in the yields of both SELECT clauses) appear in the final result:

```
FName            Age
-----------------
James            12
Joyce            11
```

Now consider the EXCEPT operator:

```
PROC SQL;
SELECT       fname, age
FROM         one
EXCEPT
SELECT       fname, age
FROM         two
;
QUIT;
```

EXCEPT is exclusive in a different way; the end result consists of rows that occur in the first operand but **not** in the second:

```
FName            Age
-----------------
Louise           12
Robert           12
Thomas           11
```

Tip: Perhaps you've noticed that the UNION, INTERSECT, and EXCEPT results all present the names in alphabetical order. That's a side effect of the processing done to derive the results. In the absence of an ORDER BY clause, the SQL processor can deliver results in any order it finds convenient.

Finally, there is the OUTER UNION operator. When we run:

```
PROC SQL;
SELECT       fname, age
FROM         one
OUTER UNION
SELECT       fname, age
FROM         two
;
QUIT;
```

we get this result:

```
FName          Age  FName          Age
----------------------------------------
Thomas          11                   .
Joyce           11                   .
James           12                   .
Robert          12                   .
Louise          12                   .
                 .   Joyce          11
                 .   James          12
                 .   John           12
                 .   Jane           12
```

We see that OUTER UNION is a most inclusive operator. It includes all of the columns and makes no attempt at consolidation. All of the rows provided by the two SELECT clauses, even the duplicate rows, are kept. Notice that the result has two columns named FNAME and two named AGE. This situation is not unusual with the OUTER UNION operator. SQL can manage repeating (nonunique) column names, but when we attempt to store the results via a CREATE TABLE statement, SAS rules (as distinguished from SQL rules) come into play and there is trouble. To demonstrate, we can run:

```
PROC SQL;
CREATE TABLE outer_union AS
SELECT      fname, age
FROM        one
OUTER UNION
SELECT      fname, age
FROM        two
;
QUIT;
```

The log reports:

```
WARNING: Variable FName already exists on file
         WORK.OUTER_UNION.
WARNING: Variable Age already exists on file
         WORK.OUTER_UNION.
NOTE: Table WORK.OUTER_UNION created, with 9 rows and 2
      columns.
```

The table looks like Exhibit 6-3.

Exhibit 6-3 OUTER_UNION

FName	Age
Thomas	11
Joyce	11
James	12
Robert	12
Louise	12
.	
.	
.	
.	

In each instance of like-named columns colliding, the first (leftmost) column prevails; others are simply discarded. However, all of the rows are kept.

6.3 Concatenation and Interleaving with OUTER UNION

We've said nothing so far about parallels between SQL set operators and DATA step techniques. That's because not all of the PROC SQL set operators have DATA step counterparts, and in some cases the DATA step counterparts are rather convoluted, or only work under restricted conditions.

However, the OUTER UNION operator, with the CORRESPONDING option in effect, does have a straightforward DATA step parallel. We can demonstrate that using the same tables (ONE and TWO) we used for the last set of examples.

The two can be combined vertically (concatenated), in a DATA step by naming them both in a single SET statement, as in:

```
DATA concat;
SET one
    two
    ;
RUN;
```

The result looks like Exhibit 6-4.

Exhibit 6-4 CONCAT (from DATA step)

FName	Weight	Age	Height
Thomas	85.0	11	.
Joyce	50.5	11	.
James	83.0	12	.
Robert	128.0	12	.
Louise	77.0	12	.
Joyce	.	11	51.3
James	.	12	57.3
John	.	12	59.0
Jane	.	12	59.8

The DATA step automatically combines data from like-named variables (FNAME and AGE in this case). When, as with WEIGHT and HEIGHT, a variable turns up in one input but not the other, missing values arise to complete the grid.

We saw earlier that the OUTER UNION operator by default does not align or consolidate columns from the two operands. That's where the CORRESPONDING option comes in. It modifies this behavior by directing the SQL processor to align like-named columns.

That's what the DATA step just did for us, so we should be able to get the same results from SQL by running:

```
PROC SQL;
CREATE TABLE concat AS
SELECT      *
FROM        one
OUTER UNION CORRESPONDING
SELECT      *
FROM        two
;
QUIT;
```

The new table (CONCAT) is shown in Exhibit 6-5.

Exhibit 6-5 CONCAT (from PROC SQL)

FName	Weight	Age	Height
Thomas	85.0	11	.
Joyce	50.5	11	.
James	83.0	12	.
Robert	128.0	12	.
Louise	77.0	12	.
Joyce	.	11	51.3
James	.	12	57.3
John	.	12	59.0
Jane	.	12	59.8

Indeed it is the same as the one generated by the DATA step. The order of the rows is even the same, although since there is no ORDER BY clause in the SQL code, the SQL processor is not obligated to deliver its results in any particular order.

The issue of row ordering also comes up if we attempt to find an SQL counterpart to an interleaving DATA step. Interleaving produces the same results as concatenation except for ordering of rows, and is implemented in the DATA step by including a BY statement after the SET statement. ONE and TWO are already sorted by AGE, so we can interleave by AGE:

```
DATA interleave;
SET one
    two
    ;
BY age;
RUN;
```

This produces the table displayed in Exhibit 6-6.

Exhibit 6-6 INTERLEAVE (from DATA step)

FName	Weight	Age	Height
Thomas	85.0	11	.
Joyce	50.5	11	.
Joyce	.	11	51.3
James	83.0	12	.
Robert	128.0	12	.
Louise	77.0	12	.
James	.	12	57.3
John	.	12	59.0
Jane	.	12	59.8

The table is the same as the one produced by concatenation, except for the ordering of the rows. The observations are now grouped by AGE, the BY variable. Within each AGE group, observations are separated by source (ONE versus TWO) with original order preserved **within** each such subgroup; this arrangement is characteristic of interleaving. It's not so easy to emulate this process with SQL. We can try adding an ORDER BY clause to our concatenation code, giving us:

```
PROC SQL;
CREATE TABLE interleave AS
SELECT      *
FROM        one
OUTER UNION CORRESPONDING
SELECT      *
FROM        two
ORDER BY    age
;
QUIT;
```

That generates the table we see in Exhibit 6-7.

Exhibit 6-7 INTERLEAVE (from PROC SQL)

FName	Weight	Age	Height
Thomas	85.0	11	.
Joyce	.	11	51.3
Joyce	50.5	11	.
James	.	12	57.3
James	83.0	12	.
Louise	77.0	12	.
Jane	.	12	59.8
Robert	128.0	12	.
John	.	12	59.0

The rows are arranged by AGE, but they are not subgrouped by source because there is
no way to specify that. We can accomplish that much by constructing queries that assign
sequence numbers, like this query:

```
PROC SQL;
SELECT      *, 1 AS Suborder
FROM        one
;
QUIT;
```

which generates this:

```
FName         Weight       Age   Suborder
------------------------------------------
Thomas            85        11          1
Joyce           50.5        11          1
James             83        12          1
Robert           128        12          1
Louise            77        12          1
```

and this query:

```
PROC SQL;
SELECT      *, 2 AS Suborder
FROM        two
;
QUIT;
```

which gives us this:

```
FName            Age    Height   Suborder
------------------------------------------
Joyce            11      51.3        2
James            12      57.3        2
John             12        59        2
Jane             12      59.8        2
```

Then we can make use of the added column (SUBORDER), before using a data set option to eliminate it. The code is:

```
PROC SQL;
CREATE TABLE interleave(DROP=suborder) AS
SELECT        *, 1 AS suborder
FROM          one
OUTER UNION CORRESPONDING
SELECT        *, 2 AS suborder
FROM          two
ORDER BY      age, suborder
;
QUIT;
```

The resulting table is as shown in Exhibit 6-8.

Exhibit 6-8 INTERLEAVE

FName	Weight	Age	Height
Joyce	50.5	11	.
Thomas	85.0	11	.
Joyce	.	11	51.3
James	83.0	12	.
Louise	77.0	12	.
Robert	128.0	12	.
James	.	12	57.3
Jane	.	12	59.8
John	.	12	59.0

The AGE groupings and source-based subgroupings are correct, but the ordering within subgroups is not preserved (for example, now Robert follows Louise, even though Louise follows Robert in the source table, ONE). That is a side effect of the sorting induced by the ORDER BY clause. In general, there is no SQL technique that precisely emulates DATA step interleaving.

Tip: If you design tables to include sequence numbers (1, 2, 3, ...) for individual rows, you can use them to gain control over ordering of rows in SQL results.

6.4 Data Type Compatibility

The alignment of columns in these examples has worked smoothly because the aligned columns have matched with respect to data type (numeric or character). Because column alignment is an essential aspect of set operators, it's worth exploring this a bit more. We need some test data sets having deliberate type mismatches, so we run:

```
DATA num;
ProblemVar = 123;
RUN;
DATA char;
ProblemVar = 'abc';
RUN;
```

Notice that the data type of PROBLEMVAR is **numeric** in data set NUM, but **character** in data set CHAR. So when we attempt a DATA step concatenation with:

```
DATA both;
SET num char;
RUN;
```

we encounter failure, with this log message:

```
ERROR: Variable ProblemVar has been defined as both
character and numeric.
```

The new data set (BOTH) is created, but contains no observations. If we run the parallel SQL code:

```
PROC SQL;
CREATE TABLE both AS
SELECT      *
FROM        num
OUTER UNION CORRESPONDING
SELECT      *
FROM        char
;
QUIT;
```

the log message is:

```
ERROR: Column 1 from the first contributor of OUTER
UNION is not the same type as its counterpart from the
second.
```

Unlike the DATA step, PROC SQL does not create even an empty table in this situation.

In attempting to align and consolidate columns, set operators assume that there is data type compatibility. As we've just seen, when this assumption is violated, the consequences are unfortunate. So, when set operators are used, the programmer must make sure that aligned columns are compatible with respect to type.

6.5 Overview: UNION, INTERSECT, and EXCEPT

With the exception of the preview presented near the start of the chapter, we've been working only with the OUTER UNION set operator and ignoring the other three operators (UNION, INTERSECT, and EXCEPT). There are two reasons for this. First, the OUTER UNION is the one set operator that has a direct parallel in the DATA step. Second, there are more differences between the OUTER UNION operator and the other three operators than there are among the other three operators.

The behavior of any of the set operators can be explained in the answers to four questions:

1. How are columns aligned?

2. What is done with columns that are left over (that is, do not align)?

3. How are rows accreted?

4. Are duplicate rows allowed in the result?

Question #3 (dealing with the rule for accreting rows) is the one we addressed in the preview (see Section 6.2). We saw that OUTER UNION and UNION behave similarly, and that INTERSECT and EXCEPT differ in distinct ways. When it comes to the other three questions, OUTER UNION stands apart and the other three operators (UNION, INTERSECT, and EXCEPT) behave exactly alike.

Column Alignment

Recall (from Section 6.3) that OUTER UNION aligns columns by name if the CORRESPONDING option is coded. The other three set operators share this feature. However, in the absence of the CORRESPONDING option, OUTER UNION does no alignment; in contrast, the default rule for UNION, INTERSECT, and EXCEPT is to align by position. In other words, the leftmost or first-named column from the first operand (SELECT clause) is aligned with the leftmost column from the second operand.

Leftover Columns

We saw that the OUTER UNION operator carries all columns forward into the result that it produces, including even those columns that cannot be matched by name when the CORRESPONDING option is used. That is the essence of "outerness."

UNION, INTERSECT, and EXCEPT are different, and the behavior depends upon whether the CORRESPONDING option is specified. If CORRESPONDING is in effect, leftover columns (those that do not match by name) from either operand are shed and do not appear in the result. If CORRESPONDING is **not** in effect, and if one operand supplies more columns than the other, the extra columns are included in the result and nulls (missing values) arise where necessary.

Row Accretion

UNION, like OUTER UNION, accepts those rows that appear in either operand (that is, in the results produced by either SELECT clause). INTERSECT accepts those rows that appear in both operands. EXCEPT accepts rows that appear in the first operand but are absent in the second.

Duplicate Rows

OUTER UNION results preserve all rows. The UNION, INTERSECT, and EXCEPT operators by default purge duplicate rows (although the optional ALL keyword can be used to prevent this).

Conclusion

Taken together, the shared characteristics of the UNION, INTERSECT, and EXCEPT set operators limit the extent to which equivalent processes can be simply coded using the DATA step. This is another point of contrast with the OUTER UNION operator.

In the sections that follow, the behavior of the UNION, INTERSECT, and EXCEPT set operators is illustrated through examples. Because UNION is probably the most widely used and shares many similarities with the others, it is covered first and most extensively. Then the distinctive characteristics of INTERSECT and EXCEPT are presented.

6.6 UNION

We begin looking at the UNION operator by using both the ALL and CORRESPONDING options. This yields the form of UNION that most closely resembles the OUTER UNION CORRESPONDING that we examined earlier. To demonstrate, still using as our data sources tables ONE (Exhibit 6-1) and TWO (Exhibit 6-2), we run:

```
PROC SQL;
CREATE TABLE unionallcorr AS
SELECT      *
FROM        one
UNION ALL CORRESPONDING
SELECT      *
FROM        two
;
QUIT;
```

This yields the table shown in Exhibit 6-9.

Exhibit 6-9 UNIONALLCORR

FName	Age
Thomas	11
Joyce	11
James	12
Robert	12
Louise	12
Joyce	11
James	12
John	12
Jane	12

Tables ONE and TWO have columns FNAME and AGE in common, so those are the columns that emerge in this result. Note that the data from the two AGE columns are properly combined in a single column, even though AGE is the third column in ONE and the second column in TWO. Each source also had an additional column (WEIGHT in ONE and HEIGHT in TWO), but these are eliminated by the UNION operator because their names do not match; that is the rule when CORRESPONDING is specified.

The ALL keyword prevents the UNION operator from eliminating duplicate rows. That's why we have a row for 11-year-old Joyce from ONE and another from TWO, and similar repetition for 12-year-old James. If we omitted ALL, we would not see such duplication.

Tip: If the ALL option were not coded in this example, the rows would be ordered differently, as a side effect of the process that detects duplicates. Generally, when you know that there are no duplicate rows, coding ALL can speed up processing by avoiding the search for duplicates.

Next, let's eliminate the CORRESPONDING option and investigate the alternative column alignment rule that then takes effect. Here is the code:

```
PROC SQL;
CREATE TABLE unionall AS
SELECT      *
FROM        one
UNION ALL
SELECT      *
FROM        two
;
QUIT;
```

The result is reflected in Exhibit 6-10.

Exhibit 6-10 UNIONALL (wrong)

FName	Weight	Age
Thomas	85.0	11.0
Joyce	50.5	11.0
James	83.0	12.0
Robert	128.0	12.0
Louise	77.0	12.0
Joyce	11.0	51.3
James	12.0	57.3
John	12.0	59.0
Jane	12.0	59.8

It would appear that there are a number of implausibly lightweight, middle-aged students. What has happened, of course, is that in the absence of the CORRESPONDING option, the columns were aligned by position rather than by name (recall that the second column in table TWO is AGE and the third column is HEIGHT).

Don't conclude that omitting the CORRESPONDING keyword always leads to trouble. That was the case here because the column naming was consistent, whereas the column ordering was not. In other situations the opposites might be true. Whenever the asterisk (*) is used in either or both of the SELECT clauses, the column alignment is to some extent implicit, and the appropriateness of the result will depend on consistency of table organization. Remember that you can always use explicit SELECT lists to control precisely the column alignment. For example, the last example could be fixed by changing the code to:

```
PROC SQL;
CREATE TABLE unionall AS
SELECT      fname, age
FROM        one
UNION ALL
SELECT      fname, age, height
FROM        two
;
QUIT;
```

That gives us the table shown in Exhibit 6-11.

Exhibit 6-11 UNIONALL (right)

FName	Age	Height
Thomas	11	.
Joyce	11	.
James	12	.
Robert	12	.
Louise	12	.
Joyce	11	51.3
James	12	57.3
John	12	59.0
Jane	12	59.8

FNAME is specified first in each operand, and AGE second; consequently, we get appropriate alignment for those two columns. The HEIGHT column emerges third, even though the first operand has only two columns. The log tells us:

```
WARNING: A table has been extended with null columns to
perform the UNION ALL set operation.
```

Tip: A SELECT list can include constants and expressions in addition to column names (see Section 2.5), so we could avoid the warning message about null columns by appending ", . AS Height" to the first SELECT list.

The alignment of columns by position has no counterpart in the DATA step. When the DATA step's SET statement is used to concatenate or interleave data sets, variables originating in different data sets are aligned strictly by name. The DATA step also lacks a mechanism for automatically eliminating variables that do not align. Instead, by default, all variables survive, with missing values arising where source data sets do not supply values. All this is another way of saying, again, that the behavior of the DATA step parallels that of PROC SQL's OUTER UNION operator with the CORRESPONDING option, and not any flavor of the simple UNION operator. This is not to say that it's impossible to develop DATA step code that emulates UNION's column alignment rules, but such code is likely to be intricate, involving complexities such as systematic renaming of variables. Because this book is primarily about SQL, we do not pursue that endeavor.

At this point we have pretty much covered column alignment issues. On the other hand, the example we have been using does not fully illustrate the issues and exercise the features pertaining to row accretion. So now we introduce a new example. It involves two tables; the first is named ABC and looks like Exhibit 6-12.

Exhibit 6-12 ABC

ID	Code
1	aa
1	aa
1	bb
1	bb
1	bb
1	bb
1	cc
1	cc

The second is named AB and looks like Exhibit 6-13.

Exhibit 6-13 AB

ID	Code
1	aa
1	aa
1	aa
1	bb
1	bb

These two tables can be generated with this code:

```
DATA ABC;
RETAIN ID 1;
DO CODE = 'aa','aa',
          'bb','bb','bb','bb',
          'cc','cc';
   OUTPUT;
   END;
RUN;
DATA ab;
RETAIN ID 1;
DO CODE = 'aa','aa','aa',
          'bb','bb';
   OUTPUT;
   END;
RUN;
```

Because the two tables have columns that agree in name, type, and position, we can use the asterisk shorthand notation in the SELECT list and know that the columns will align appropriately, whether or not the CORRESPONDING option is invoked. In other words, we have a special case in which column alignment is not an issue. That helps us, in the following examples, to focus on the issues surrounding row accretion. It also allows us to construct DATA steps that are equivalent to the SQL set operators (something we've already concluded to be highly complicated when column alignment is an issue).

Notice that tables ABC and AB are already sorted. That was done just to simplify DATA step examples by eliminating the need for PROC SORT steps. For PROC SQL, the ordering of the rows does not matter.

Let's start the exploration of row accretion by examining the effect of the ALL option. It's a negative option, in the sense that coding it causes PROC SQL to **not** do something (purge duplicates) that it would otherwise do by default. So the query:

```
PROC SQL;
CREATE TABLE unionall AS
SELECT        *
FROM          abc
UNION ALL
SELECT        *
FROM ab
;
QUIT;
```

has as its result the table seen in Exhibit 6-14.

Exhibit 6-14 UNIONALL (from PROC SQL)

ID	Code
1	Aa
1	Aa
1	Bb
1	Bb
1	Bb
1	Bb
1	Cc
1	Cc
1	Aa
1	Aa
1	Aa
1	Bb
1	Bb

This is simply the concatenation of the two sources (tables ABC and AB). Of course, the ordering is incidental, since no ORDER BY clause was coded. The significant thing is the number of times each distinct row appears. The accretion rule for the UNION operators is that a row appears in the result if it appears in **either** source. When the ALL option is used, the number of times it appears is the sum of its populations in the two sources. That is, if F represents the number of times a distinct row appears in the first source (the result of the first SELECT clause) and S represents the count from the second source, the row will appear F+S times in the result. Thus, because the 1/bb pair appears four times in ABC and twice in AB, it appears six times in the UNION. Note that this is also the row accretion rule used by the OUTER UNION operator.

Now let's see what happens when the ALL keyword is removed. If we run:

```
PROC SQL;
CREATE TABLE union AS
SELECT      *
FROM        abc
UNION
SELECT      *
FROM        ab
;
QUIT;
```

the result is the table seen in Exhibit 6-15.

Exhibit 6-15 UNION (from PROC SQL)

ID	Code
1	aa
1	bb
1	cc

As stated earlier, in the absence of the ALL option, the duplicate rows are purged. This process is performed **after** the rows from the two sources are pooled (if it were performed before, there could still be pairs of identical rows in the result).

The UNION operator is commutative, meaning that the results are not changed (except for row order) if the two operands are interchanged. However, in some situations (though not in our present example) column names and other attributes could be affected by such a switch.

We've already concluded that there is no reasonably simple way to fully emulate the UNION operator using the DATA step, because the column alignment rules are different. But if we focus just on the matter of row accretion, we can certainly achieve the simple concatenation, as in this code:

```
DATA unionall;
SET abc ab;
RUN;
```

which produces the same result as the SQL set operator UNION ALL (see Exhibit 6-14). Adding this PROC SORT step:

```
PROC SORT DATA=unionall OUT=union NODUPRECS;
BY _ALL_;
RUN;
```

eliminates duplicate records, thus yielding the same result as the UNION without the ALL option (see Exhibit 6-15).

Reference: Read more about PROC SORT in the *Base SAS 9.2 Procedures Guide*: Procedures: The SORT Procedure, or in *The Little SAS Book* (Fourth Edition): Section 4.3.

We can also use the MERGE statement as the basis of emulation. It is a little less obvious, but we'll illustrate it here because it is a good foundation for developing DATA step counterparts for the INTERSECT and EXCEPT operators.

Because ABC and AB are in the appropriate sort order, this DATA step:

```
DATA union;
MERGE abc(IN=in_abc) ab(IN=in_ab);
BY id code;
IF FIRST.code and (in_abc or in_ab);
RUN;
```

could be run to generate the three distinct rows that constitute the UNION result. Here is a diagram representing the process:

ABC	AB	Result
1|aa	1|aa	1|aa
1|aa	1|aa	
	1|aa	
1|bb	1|bb	1|bb
1|bb	1|bb	
1|bb		
1|bb		
1|cc		1|cc
1|cc		

Note that each cell in this diagram represents a row of data.

The UNION ALL is a bit harder to emulate, but it can be done with this code:

```
DATA unionall;
in_abc = 0;
in_ab  = 0;
MERGE abc(IN=in_abc) ab(IN=in_ab);
BY id code;
IF in_abc THEN OUTPUT;
IF in_ab  THEN OUTPUT;
RUN;
```

The two assignment statements manipulate the variables that are otherwise maintained by the IN= data set options. This is somewhat advanced DATA step programming, and tracking through the details is beyond the scope of this book. However, we can consider this diagram:

ABC	AB	Result	
1|aa	1|aa	1|aa	1|aa
1|aa	1|aa	1|aa	1|aa
*	1|aa		1|aa
1|bb	1|bb	1|bb	1|bb
1|bb	1|bb	1|bb	1|bb
1|bb	*	1|bb	
1|bb	*	1|bb	
1|cc		1|cc	
1|cc		1|cc	

The asterisks represent places that would have given rise to "phantom" rows had we not explicitly reinitialized the IN= flags. There are two groups of result rows, corresponding to the two OUTPUT statements. The data set (UNIONALL) that emerges is shown in Exhibit 6-16.

Exhibit 6-16 UNIONALL (from DATA step using MERGE)

ID	Code
1	aa
1	aa
1	aa
1	aa
1	aa
1	bb
1	bb
1	bb
1	bb
1	bb
1	bb
1	cc
1	cc

It is the same as the SQL UNION ALL result, except for the order of the rows. That's not surprising, since we're using a merge process here, whereas the SQL UNION does a simple concatenation. It's as close as we can come without making the code even more complicated.

6.7 INTERSECT

We turn now to the INTERSECT operator. With regard to column alignment, it behaves precisely as the UNION operator does, so we don't repeat those details here. Instead, we concentrate on the row accretion process, where we find a distinction: whereas the UNION operator accepts rows that appear in **either** source, INTERSECT accepts only those rows that appear in **both** sources. Once again, we demonstrate first with the ALL option in effect, as in this code:

```
PROC SQL;
CREATE TABLE intersectall AS
SELECT        *
FROM          abc
INTERSECT ALL
SELECT        *
FROM          ab
;
QUIT;
```

Before looking at the result, let's try to sketch what the SQL processor has been asked to do.

ABC	AB	Result
1|aa	1|aa	1|aa
1|aa	1|aa	1|aa
	1|aa	
1|bb	1|bb	1|bb
1|bb	1|bb	1|bb
1|bb		
1|bb		
1|cc		
1|cc		

Basically, each row that appears in the first operand and again in the second becomes part of the result. Moreover, such pairs can repeat, placing duplicate rows in the result. Specifically, if F represents the number of times a particular row appears in the first source (the result of the first SELECT clause) and S represents the count from the second source, the row will appear MIN(F,S) times in the result.

So the output table is as shown in Exhibit 6-17.

Exhibit 6-17 INTERSECTALL

ID	Code
1	aa
1	aa
1	bb
1	bb

Our diagram also suggests how a DATA step might be coded to yield the same result:

```
DATA intersectall;
in_abc = 0;
in_ab  = 0;
MERGE abc(IN=in_abc) ab(IN=in_ab);
BY id code;
IF in_abc AND in_ab;
RUN;
```

Turning back to SQL and the INTERSECT operator, if we remove the ALL option, leaving the query as:

```
PROC SQL;
CREATE TABLE intersect AS
SELECT      *
FROM        abc
INTERSECT
SELECT      *
FROM        ab
;
QUIT;
```

the duplicates are removed and the result is what we see in Exhibit 6-18.

Exhibit 6-18 INTERSECT

ID	Code
1	aa
1	bb

Our diagram of the process looks like this:

ABC	AB	Result
1|aa	1|aa	1|aa
1|aa	1|aa	
	1|aa	
1|bb	1|bb	1|bb
1|bb	1|bb	
1|bb		
1|bb		
1|cc		
1|cc		

Here's the DATA step that produces the same output:

```
DATA intersect;
MERGE abc(IN=in_abc) ab(IN=in_ab);
BY id code;
IF FIRST.code AND in_abc AND in_ab;
RUN;
```

Like the UNION operator, INTERSECT is commutative. The positions of the operands can be switched without affecting the content of the result.

6.8 EXCEPT

Finally, we consider the EXCEPT operator. With regard to column alignment, it too behaves just as the UNION operator does, so we don't repeat those details here. The EXCEPT operator's accretion rule is to preserve any row that appears in the first operand (SELECT clause), but is **not** matched in the second. Another way of saying this is that rows are taken from the first operand unless they are "canceled" by virtue of appearance in the second operand. To illustrate first with the ALL option in effect, this code:

```
PROC SQL;
CREATE TABLE exceptall AS
SELECT      *
FROM        abc
EXCEPT ALL
SELECT      *
FROM        ab
;
QUIT;
```

gives us the result shown in Exhibit 6-19.

Exhibit 6-19 EXCEPTALL

ID	Code
1	bb
1	bb
1	cc
1	cc

The diagram looks like this:

ABC	AB	Result
1|aa	1|aa	
1|aa	1|aa	
	1|aa	
1|bb	1|bb	
1|bb	1|bb	
1|bb		1|bb
1|bb		1|bb
1|cc		1|cc
1|cc		1|cc

If F represents the number of times a particular row appears in the first source (the result of the first SELECT clause) and S represents the count from the second source, the row will appear MAX(0,F-S) times in the result.

Once again, we can adapt our DATA step to yield the same result. The code is now:

```
DATA exceptall;
in_abc = 0;
in_ab  = 0;
MERGE abc(IN=in_abc) ab(IN=in_ab);
BY id code;
IF in_abc AND NOT in_ab;
RUN;
```

If we remove the ALL option from the SQL code, leaving the query as:

```
PROC SQL;
CREATE TABLE except AS
SELECT        *
FROM          abc
EXCEPT
SELECT        *
FROM          ab
;
QUIT;
```

the result is the table displayed in Exhibit 6-20.

Exhibit 6-20 EXCEPT

ID	Code
1	cc

The duplicates are removed **before** rows from the two sources are paired and canceled. So the diagram is a bit more involved than in the other situations we have represented. It looks like this:

The processor first removes duplicate rows **within** each operand. The result of that step is represented in the center of the diagram. Then the matching is done.

The equivalent DATA step code is:

```
DATA except;
MERGE abc(IN=in_abc) ab(IN=in_ab);
BY id code;
IF FIRST.code AND in_abc AND NOT in_ab;
RUN;
```

Unlike UNION and INTERSECT, EXCEPT is **not** commutative. Switching the operands changes the result. To illustrate, we run:

```
PROC SQL;
SELECT      *
FROM        ab
EXCEPT ALL
SELECT      *
FROM        abc
;
QUIT;
```

which gives us:

```
     ID  Code
-------------
      1  aa
```

Without the ALL option, the result set would be empty (zero rows).

6.9 Summary

Set operators complement joins and subqueries by providing other ways of combining data from multiple sources. Typically, set operators perform end-to-end (vertical) combinations, in contrast to the side-by-side (horizontal) combinations that result from joins.

The OUTER UNION operator in a number of ways resembles the operation of a SET statement that processes two data sets in a DATA step. The other three set operators (UNION, INTERSECT, and EXCEPT) differ in nature from the OUTER UNION operator. They also differ from each other in terms of the set-theoretic rules they implement, but resemble one another in terms of their mechanics. UNION, INTERSECT, and EXCEPT do not have simple DATA step counterparts, though in special cases some emulation can be programmed.

Chapter 7

Global Statements, Options, and Session Management

In the preceding chapters, we explored most of the features of the SELECT statement. One thing we have not yet discussed is the ways in which their behavior is conditioned by global statements and by SQL procedure options.

Global statements are those that can appear either within a SAS step or between steps. In either case, they take effect immediately and remain in effect for the duration of the job or session, unless overridden by subsequent global statements. The TITLE statement is a good, commonplace example. Global statements typically have the same effects on PROC SQL as they have on other parts of SAS, so we don't bother to enumerate or discuss them individually. What is different, however, is the way their position in the code stream within a PROC SQL step affects their interaction with SQL statements; that's the one aspect of global statements we examine in this chapter.

> **Reference:** Read more about global statements in the *SAS 9.2 Language Reference: Dictionary*: Dictionary of Language Elements: Statements: Global Statements.

SAS has at least four types of options:

- system options, which, unless explicitly changed or nullified, remain in effect for the duration of a job or session
- data set options, which are syntactically bound to data set references
- procedure options, which are typically coded on the PROC statement and apply to the procedure step as an entity
- statement-specific options

The different scopes of the four types of options can make things a little more confusing. System options and data set options operate with PROC SQL pretty much as they do with other parts of SAS, so we do not take them up here. PROC SQL statements have no statement-specific options.

> **Reference:** Read more about system options and data set options in *The Little SAS Book* (Fourth Edition): Sections 1.14 (system options) and 6.11 (data set options).

That leaves us with the procedure options as a topic for consideration in this chapter. We don't enumerate and explain them; that's a job better left to the documentation. We do, however, illustrate their use and management.

> **Reference:** The available PROC SQL options are listed and explained in the *Base SAS 9.2 Procedures Guide*: Procedures: The SQL Procedure: Syntax: SQL Procedure: PROC SQL Statement.

Recall that PROC SQL, unlike most parts of SAS, completely processes each statement (such as SELECT or CREATE) before even beginning to analyze the next statement. That characteristic has important implications for global statements that appear within PROC SQL steps, and for PROC SQL options.

In the previous chapters, most of the SQL code examples revolve around single statements (usually SELECT or CREATE statements), each preceded by a PROC SQL statement and followed by a QUIT statement. That structure makes each example self-contained and suitable to be run, as is. That structure may also create the false impression that SELECT and CREATE and other statements **must** be isolated in separate steps. That is not the case, and in this chapter we see a number of examples in which a single PROC SQL step comprises numerous statements.

Before proceeding to the details, let's set up a small table to use in the examples by running this code:

```
PROC SQL;
CREATE TABLE twelves AS
SELECT      name as FName,
            sex,
            height,
            weight
FROM        sashelp.class
WHERE       age=12
;
QUIT;
```

The table looks like Exhibit 7-1.

Exhibit 7-1 TWELVES

FName	Sex	Height	Weight
James	M	57.3	83.0
Jane	F	59.8	84.5
John	M	59.0	99.5
Louise	F	56.3	77.0
Robert	M	64.8	128.0

7.1 Global Statements

The TITLE statement is a widely used global statement and serves as a good example. Before considering its usage with PROC SQL, let's examine its usage with another procedure, as in:

```
PROC TABULATE DATA=twelves NOSEPS;
CLASS sex;
VAR height weight;

TITLE 'Weight Minima by Sex';
TABLE sex, weight * MIN * F=7.1;
```

```
TITLE 'Height Maxima by Sex';
TABLE sex, height * MAX * F=7.1;

RUN;
```

Intuitively, you might expect to see two small tables in the output, each with an appropriate title. In fact, the output looks like this:

```
Height Maxima by Sex
------------------------------
|                    |Weight |
|                    |-------|
|                    |  Min  |
|--------------------+-------|
|Sex                 |       |
|F                   |   77.0|
|M                   |   83.0|
------------------------------
Height Maxima by Sex
------------------------------
|                    |Height |
|                    |-------|
|                    |  Max  |
|--------------------+-------|
|Sex                 |       |
|F                   |   59.8|
|M                   |   64.8|
------------------------------
```

Notice that the second title is used, incorrectly, for the first table. That happens because SAS detects all of the global statements coded within or immediately before the step and puts them into effect **before** the step begins its work. So the preceding code is essentially equivalent to the following:

```
TITLE 'Weight Minima by Sex';
TITLE 'Height Maxima by Sex';

PROC TABULATE DATA=twelves NOSEPS;
CLASS sex;
VAR height weight;

TABLE sex, weight * MIN * F=7.1;

TABLE sex, height * MAX * F=7.1;

RUN;
```

Now it's rather obvious that the first TITLE statement is overridden before it is ever used. To have a different TITLE statement in effect for each TABLE statement, you would have to code two separate PROC TABULATE steps, like this:

```
TITLE 'Weight Minima by Sex';
PROC TABULATE DATA=twelves NOSEPS;
CLASS sex;
VAR height weight;
TABLE sex, weight * MIN * F=7.1;
RUN;
TITLE 'Height Maxima by Sex';
PROC TABULATE DATA=twelves NOSEPS;
CLASS sex;
VAR height weight;
TABLE sex, height * MAX * F=7.1;
RUN;
```

Now consider SQL code that produces similar output:

```
PROC SQL;

TITLE 'Weight Minima by Sex';
SELECT      sex,
            MIN(weight) FORMAT=7.1 LABEL='Weight Min'
FROM        twelvesGROUP BY      sex
;

TITLE 'Height Maxima by Sex';
SELECT      sex,
            MAX(height) FORMAT=7.1 LABEL='Height Max'
FROM        twelvesGROUP BY      sex
;

QUIT;
```

The result is:

```
Weight Minima by Sex

      Weight
Sex      Min
------------
F       77.0
M       83.0
```

```
Height Maxima by Sex

        Height
Sex       Max
-----------
F        59.8
M        64.8
```

Each little table appears under the appropriate title. That's because PROC SQL, unlike most SAS procedures, analyzes and executes one statement at a time before proceeding to the next statement. You can intersperse global statements with the SQL statements, with the effects occurring in the order in which the statements are coded.

7.2 PROC SQL Options

A number of options can be specified within the PROC SQL statement and remain in effect for the duration of the SQL job or session (or, as we'll see a bit later, until nullified or changed via the RESET statement). Most can be grouped into a few broad categories.

- Several options are useful in debugging and performance tuning. A couple of examples appear later in this chapter.

- One option in particular, PRINT|NOPRINT, is useful in conjunction with SQL code that creates and populates macro variables. We will discuss it when we get to that topic (see Section 8.2).

- Several options are designed to enhance PROC SQL's usefulness as a report generator (that is, as a tool to display results itself, via a stand-alone SELECT statement, rather than to store results in a table). See Chapter 11.

Reference: Read more about PROC SQL options in the SAS documentation. For a complete list of the supported options, see the *Base SAS 9.2 Procedures Guide*: Procedures: The SQL Procedure: Syntax: SQL Procedure: PROC SQL Statement.

Now let's demonstrate the use of a few of these options. Let's display the information on the 12-year-old students, listed in order of height from shortest to tallest. We use the FEEDBACK option to have the log display the SQL code after it has been standardized and expanded by the processor. We use the STIMER option to display the amount of time used to process each statement. So the code looks like this:

```
PROC SQL      FEEDBACK STIMER;
SELECT        *
FROM          twelves
ORDER BY      height
  ;
```

Note the absence of a QUIT statement here. PROC SQL processes the SELECT statement and continues running. The output looks like this:

```
FName      Sex     Height     Weight
------------------------------------
Louise     F         56.3         77
James      M         57.3         83
John       M           59       99.5
Jane       F         59.8       84.5
Robert     M         64.8        128
```

In the log we see, courtesy of the FEEDBACK option:

```
NOTE: Statement transforms to:

      select TWELVES.FName, TWELVES.Sex,
             TWELVES.Height, TWELVES.Weight
        from WORK.TWELVES
    order by TWELVES.Height asc;
```

and, from the STIMER option:

```
NOTE: SQL Statement used (Total process time):
      real time          0.07 seconds
      cpu time           0.01 seconds
```

Now suppose that we only want to see the rows for the three shortest students and that we no longer want to see notes produced by the STIMER option. The OUTOBS= option can do the subsetting. So we could use a QUIT statement to end the SQL session, followed by a PROC SQL statement to relaunch SQL with the options we want:

```
PROC SQL      FEEDBACK OUTOBS=3;
```

We would follow this with the SELECT statement. However, we don't have to do all that. Instead, we can take advantage of the RESET statement, whose purpose is to allow SQL procedure options to be invoked, negated, or changed in the midst of a PROC SQL step. So the code for the entire step is:

```
PROC SQL       FEEDBACK STIMER;

SELECT         *
FROM           twelves
ORDER BY       height
;
RESET          NOSTIMER OUTOBS=3;
SELECT         *
FROM           twelves
ORDER BY       height
;

QUIT;
```

The FEEDBACK option is in effect for both SELECT statements, whereas STIMER applies only to the first SELECT statement and OUTOBS=3 applies only to the second SELECT statement. The output from the first SELECT is unchanged from what we saw earlier. The output from the second SELECT, as expected, looks like this:

```
FName      Sex    Height    Weight
-----------------------------------
Louise     F        56.3        77
James      M        57.3        83
John       M          59      99.5
```

The RESET statement clearly provides a lot of flexibility. However, it can also make code difficult to trace and analyze, especially in a long PROC SQL step having multiple RESET statements, each of which toggles or specifies a different group of options. You might find it helpful to abide by some self-imposed limitations:

- Options that are to remain in effect for the entire session or job should be coded on the PROC SQL statement.
- Options that are to apply to just one SQL statement (or to a group of consecutive statements) should be invoked using a RESET statement at the point where they are needed, and the same group of options should be reversed by another RESET statement at the point where they are no longer needed.

Following these rules, we would change our code to:

```
PROC SQL      FEEDBACK;

RESET         STIMER;
SELECT        *
FROM          twelves
ORDER BY      height
;
RESET         NOSTIMER;

RESET         OUTOBS=3;
SELECT        *
FROM          twelves
ORDER BY      height
;
RESET         OUTOBS=MAX;

QUIT;
```

7.3 Summary

SAS global statements and PROC SQL options can be used to affect the behavior of PROC SQL. They can also be specified or respecified between SQL statements, with such changes taking effect immediately.

Chapter **8**

Using the Macro Facility with PROC SQL

We noted earlier that the SAS macro facility is part of the environment in which PROC SQL operates; in this chapter we confirm that macros can be used to generate PROC SQL code. Then we look at the opposite side of things and show how PROC SQL can create and populate macro variables for use in subsequent PROC SQL code and elsewhere.

8.1 Generating PROC SQL Code

You can use a SAS macro to generate PROC SQL code. This by itself is not surprising and really does not merit much discussion. After all, a macro is a tool to generate SAS code, and PROC SQL code is SAS code. It would be surprising, and would require explanation, were it **not** possible to generate PROC SQL code using the macro facility.

What is worth mentioning is that the macro facility gives us ways to work around some of the limitations of SQL. In particular, we can use a macro to compensate for the absence of array support in SQL.

To illustrate, we need a table that has a number of columns whose names differ only by a numeric suffix. So we run this code:

```
DATA wide;
INPUT ID $ Measure1-Measure4;
CARDS;
A 11 12 13 14
B 21 22 23 24
;
```

The resulting table (WIDE) looks like Exhibit 8-1.

Exhibit 8-1 WIDE

ID	Measure1	Measure2	Measure3	Measure4
A	11	12	13	14
B	21	22	23	24

Let's suppose that the task at hand is to add up each of the MEASURE*n* columns and store the four totals in a new data set. The simplest and most straightforward technique is to use **PROC SUMMARY**, which is designed for such tasks. If we run:

```
PROC SUMMARY DATA=wide;
VAR Measure1-Measure4;
OUTPUT OUT=sums(DROP = _type_ _freq_)
SUM = Sum1-Sum4;
RUN;
```

we get the table shown in Exhibit 8-2. This is the result we were seeking.

Exhibit 8-2 SUMS

Sum1	Sum2	Sum3	Sum4
32	34	36	38

It's also possible to do this in a DATA step. When there are many variables to be processed, the usual approach is to create arrays and to code one or more DO loops to coordinate references to the arrays. Using that approach here, we have:

```
DATA sums;
SET wide END=last;
ARRAY _measure{*} measure1-measure4;
ARRAY _sum{*} Sum1-Sum4;
KEEP sum1-sum4;
DO i = 1 TO 4;
   _sum{i} + _measure{i};
   end;
IF last THEN OUTPUT;
RUN;
```

We get, again, the table shown in Exhibit 8-2.

Now suppose we want SQL to do the same thing. We quickly encounter an obstacle because SQL has no features even remotely resembling the DATA step's ARRAY and DO statements. The fundamental reason for that is that SQL is designed to work with normalized data structures, and normalized data structures do not have column groups requiring parallel treatment. So the best-practice solution is probably to normalize the data structure and work from there.

Preview: In Section 8.2 we work with a slightly different version of this example, and in Section 12.3 we carry out the recommended change of structure.

However, there might be situations where structure change is precluded. That makes it necessary to use what can become very repetitious and verbose code. We can get the column sums by running:

```
PROC SQL;
CREATE TABLE sums AS
SELECT        SUM(measure1)  AS Sum1,
              SUM(measure2)  AS Sum2,
              SUM(measure3)  AS Sum3,
              SUM(measure4)  AS Sum4
FROM          wide
;
QUIT;
```

Once more, the result is as shown in Exhibit 8-2.

That's not so bad with four columns, but what if there were 400? That's where the macro facility can be really helpful. We can code a relatively simple macro like this:

```
%MACRO selectsums(maxindex=);
    %DO n = 1 %TO &maxindex;
                SUM(measure&n) as Sum&n
        %IF &n NE &maxindex %THEN   ,
          ;
        %END;
    %MEND selectsums;
```

and invoke it in context like this:

```
PROC SQL;
CREATE TABLE sums AS
SELECT          %selectsums(maxindex=4)
FROM            wide
;
QUIT;
```

The macro generates the correct SELECT list, and the result is, once again, reflected in Exhibit 8-2. This works because there are a number of suitably named columns to be treated in parallel. By having only four, we keep the example compact. However, the technique works just as well with many more columns, making the advantages far more dramatic.

Of course, this technique depends on the names of the columns following the rather rigid pattern that we assumed for this example. Later, when we look at DICTIONARY tables, we'll come back to this problem and develop a more versatile solution.

> **Reference:** For more information about using the SAS Macro Facility with PROC SQL, see the *SAS 9.2 SQL Procedure: User's Guide*: Programming with the SQL Procedure: Using PROC SQL with the SAS Macro Facility.

8.2 Populating Macro Variables

We have just illustrated the use of the macro facility (which itself is not part of PROC SQL) to generate PROC SQL code. Now we turn to PROC SQL features that create and populate macro variables. Before we begin exploring these capabilities, we need another table to support our examples. We can derive it from the SASHELP.CLASS table by running:

```
PROC SQL;
CREATE TABLE thirteens AS
SELECT      name AS FName,
            height FORMAT=6.1,
            weight FORMAT=6.1
FROM        sashelp.class
WHERE       age=13
;
QUIT;
```

The result (THIRTEENS) is seen in Exhibit 8-3.

Exhibit 8-3 THIRTEENS

FName	Height	Weight
Alice	56.5	84.0
Barbara	65.3	98.0
Jeffrey	62.5	84.0

User-Defined Macro Variables

In developing flexible SAS applications, it is extremely valuable to be able to programmatically create and populate macro variables. To illustrate, suppose we want to display the information on the heights of the 13-year-olds, with a footnote to provide the average height (rounded to the nearest tenth of an inch). We could use PROC MEANS to compute that average, and then type it into a FOOTNOTE statement. Fortunately, macro variables give us a way to mechanize that process. In the DATA step, it's done with the CALL SYMPUT routine, as in:

```
DATA _NULL_;
SET thirteens END=lastobs;
heightsum + height;
IF (lastobs) THEN
 CALL SYMPUT('avgheight', PUT(heightsum / _N_, 4.1) );
RUN;
```

Notice that the _NULL_ keyword is coded to preclude creation of an output data set. The only purpose of this DATA step is to accumulate a running total of HEIGHT, and at the end to transform this into an average and load it into the macro variable &AVGHEIGHT. To monitor the outcome, we can conclude with a %PUT statement like this:

```
%put macro variable AVGHEIGHT: [&avgheight];
```

to present the result in the log. The square brackets are used as a container to enable us to see whether there are leading or trailing blanks in &AVGHEIGHT. In this case we see:

```
macro variable AVGHEIGHT: [61.4]
```

Then it is a simple matter to generate the report by running:

```
TITLE 'Heights of 13-Year-Olds';
FOOTNOTE "Average Height is &avgheight";
PROC PRINT DATA=thirteens;
ID fname;
VAR height;
RUN;
```

which produces this output:

```
Heights of 13-Year-Olds

 FName      Height

Alice        56.5
Barbara      65.3
Jeffrey      62.5

Average Height is 61.4
```

Now let's do the same thing using PROC SQL. Because SQL is a non-procedural language, PROC SQL cannot support the use of CALL routines. So CALL SYMPUT is not available. However, PROC SQL has its own tool for creating and populating macro variables: the INTO clause. The INTO clause is part of the SELECT statement or clause. When the INTO clause is used, it precedes the FROM clause.

Turning to our example, we would create the macro variable by running:

```
PROC SQL;
RESET        NOPRINT;
SELECT       PUT(MEAN(height),4.1)
INTO         : avgheight
FROM         thirteens
;
RESET        PRINT;
%PUT macro variable AVGHEIGHT: [&avgheight];
```

The SELECT statement computes the required average and calls the PUT function to convert it to a character string, rounded to one decimal place. That's all courtesy of the expression that immediately follows SELECT, and does not involve anything we haven't seen before. It's the INTO clause that's new. It is followed by what is called a host variable, distinguished by a preceding colon. This construct tells the SQL processor to create a macro variable (&AVGHEIGHT) and load it with the result generated when the preceding expression is evaluated. As before, we use a %PUT statement to let us see the result in the log. It shows:

```
macro variable AVGHEIGHT: [61.4]
```

This result confirms that the SELECT statement with an INTO clause to populate the macro variable in PROC SQL is equivalent to the DATA step with the CALL SYMPUT statement.

PROC SQL would, by default, also display the result of the SELECT statement in the output window or the listing file. We toggled the PRINT | NOPRINT option to suppress that. In this example, only the SELECT statement (including the INTO and FROM clauses) is essential; the RESET and %PUT statements are present only for convenience.

With the macro variable loaded, it's simple to generate our little report. We deliberately omitted the QUIT statement after the code that created the macro variable so all we have to code here is:

```
TITLE 'Heights of 13-Year-Olds';
FOOTNOTE "Average Height is &avgheight";
SELECT        fname, height
FROM          thirteens
;
QUIT;
```

The result looks like this:

```
Heights of 13-Year-Olds

FName      Height
---------------
Alice        56.5
Barbara      65.3
Jeffrey      62.5

Average Height is 61.4
```

Tip: Macro variables generated in one PROC SQL statement can be used in later statements within the same PROC SQL step.

We have generated this little report twice, once without PROC SQL and once with PROC SQL. When we used the DATA step to create the macro variable, we used PROC PRINT to do the report. When we used SQL to create the macro variable, we used SQL to do the report. However, we could have taken a mix-and-match approach. Macro variables created by PROC SQL can be used outside of PROC SQL, and PROC SQL code can incorporate macro variables created outside of PROC SQL.

The example with which we've been working is simple in that the SELECT statement with the INTO clause yields one row and one column, and populates one macro variable. The mechanism can do more.

Suppose we have multiple items in the SELECT list, as in:

```
PROC SQL;
SELECT       MAX(fname), MIN(fname)
FROM         thirteens
;
QUIT;
```

Note that the SELECT list includes nothing but summary information, and that there is no GROUP BY clause, so we can still be sure the result will be a single row. There are two items in the SELECT list, so the result will have two columns. To create and load a macro variable for each, we simply code an INTO clause with a parallel list (that is, a list with two macro variable names). The statement becomes:

```
PROC SQL;
SELECT        MAX(fname),   MIN(fname)
INTO          : max_fname , : min_fname
FROM          thirteens
;
QUIT;
```

Notice the leading colon in each item of the INTO list. We can confirm that the macro variables have been populated by using %PUT statements to display their values:

```
%PUT macro variable MAX_FNAME: [&max_fname];
%PUT macro variable MIN_FNAME: [&min_fname];
```

In the log we see:

```
macro variable MAX_FNAME: [Jeffrey ]
macro variable MIN_FNAME: [Alice    ]
```

We were careful to make the two lists (the one following SELECT and the one following INTO) the same length. If you are careless and there is a discrepancy (either more columns than macro variables or more macro variables than columns), SAS issues a warning. Thus this code:

```
PROC SQL;
SELECT          MAX(weight), MIN(weight)
INTO            : max_weight
FROM            thirteens
;
QUIT;
```

results in:

```
WARNING: INTO clause specifies fewer host variables than
columns listed in the SELECT clause.
```

Then running:

```
%PUT macro variable MAX_WEIGHT: [&max_weight];
%PUT macro variable MIN_WEIGHT: [&min_weight];
```

yields:

```
macro variable MAX_WEIGHT: [      98]
WARNING: Apparent symbolic reference MIN_WEIGHT not
resolved.
```

This tells us that SAS has created the macro variable for the first column only. Now let's experiment with the opposite condition: providing too many macro variable names. If we run:

```
PROC SQL;
SELECT          MAX(height)
INTO            : max_height, : min_height
FROM            thirteens
;
QUIT;
```

we see:

```
WARNING: INTO clause specifies more host variables than
columns listed in the SELECT clause. Surplus host
variables will not be set.
```

When we follow up with:

```
%PUT macro variable MAX_HEIGHT: [&max_height];
%PUT macro variable MIN_HEIGHT: [&min_height];
```

we get:

```
macro variable MAX_HEIGHT: [    65.3]
WARNING: Apparent symbolic reference MIN_HEIGHT not
resolved.
```

Once again, SAS has created just the first macro variable.

Now we turn to the situation in which the SELECT statement yields more than one row. To keep things simple and focus on the issue, we use a query that yields a single column, such as:

```
PROC SQL;
SELECT       fname
FROM         thirteens
;
QUIT;
```

There are three rows in **THIRTEENS**, so the result of this query has three rows. Thus we would like to end up with three macro variables. It's pretty straightforward with the DATA step; we can run:

```
DATA _NULL_;
SET thirteens;
CALL SYMPUT('fname'||COMPRESS(PUT(_N_,4.) ),fname);
RUN;
```

When we display the results by means of this %PUT statement:

```
%PUT macro variables FNAME1 through FNAME3:
  [&fname1,&fname2,&fname3];

%PUT macro variables: [&fname1,&fname2,&fname3];
```

we see:

```
macro variables: [Alice    ,Barbara ,Jeffrey ]
```

This confirms that SAS loaded the three names into macro variables.

Now let's address the same problem using SQL. We can introduce an INTO clause like the one used in our earlier example:

```
PROC SQL;
SELECT      fname
INTO        : fname
FROM        thirteens
;
QUIT;
```

To place the result in the log, we run:

```
%PUT macro variable FNAME: [&fname];
```

The result is:

```
macro variable FNAME: [Alice   ]
```

There are three rows in the source table, but the macro variable only receives a value from the first of the three. So we need a way to create targets for the additional rows of data. We know that we cannot accomplish this by extending the INTO list with more items separated by commas; that would create a mismatch vis-à-vis the single item in the SELECT list.

PROC SQL has a different bit of syntax to allow a query to create and populate an entire vector of macro variables from a single column of data. To accomplish this, we expand the INTO item from a single macro variable to a pair of macro variables having the same name stem and appropriate integer suffixes. The two names are separated by the keyword THROUGH. As always in the INTO clause, the names of macro variables are preceded by colons.

In our example, there are three rows. So the code becomes:

```
PROC SQL;
SELECT      fname
INTO        : fname1 THROUGH : fname3
FROM        thirteens
;
QUIT;
```

If we follow this with:

```
%PUT macro variables: [&fname1,&fname2,&fname3];
```

we see:

```
macro variables: [Alice,Barbara,Jeffrey]
```

This is just like the DATA step result, except that PROC SQL has eliminated the trailing blanks. The DATA step has a seemingly important advantage, however. It does not require that the number of macro variables be hardcoded, whereas the SQL code does. Often, when you write an SQL SELECT statement that is to return multiple rows, you don't know exactly how many rows will materialize in the result. In fact, the number often changes from one run to the next as you introduce new data.

There is a simple solution, though it's a bit inelegant. We can specify a very high number of macro variables (with "very high" meaning higher than the highest number we expect to actually encounter). For example, if we know that we will never have more than a few dozen students, we might change our code to:

```
PROC SQL;
SELECT      fname
INTO        : fname1 THROUGH : fname999
FROM        thirteens
;
QUIT;
```

SAS does not make an issue of the specification of an excessive number of macro variables. If we try to reference one of the unneeded targets, as with:

```
%PUT macro variable FNAME4: [&fname4];
```

we get:

```
WARNING: Apparent symbolic reference FNAME4 not
resolved.
```

This is actually reassuring; it tells us that SAS does not create the unneeded macro variables.

We have not discussed the usefulness of such macro variable constellations. It turns out that they can be useful, but only with some fairly sophisticated macro coding techniques. Often, it is more useful to concatenate the multiple data values into a single (though perhaps lengthy) macro variable.

Suppose we want to extract all of the names and string them out in a macro variable, separated by blank-slash-blank sequences. It's not too complicated with DATA step code. Consider this code:

```
DATA _NULL_;
SET thirteens END=done;
LENGTH fname_string $ 9999;
RETAIN fname_string;
fname_string = CATX(' / ',fname_string,fname);
IF done THEN CALL SYMPUT('fnames',TRIM(fname_string) );
RUN;


%PUT FNAMES: [&fnames];
```

The result is:

```
FNAMES: [Alice / Barbara / Jeffrey]
```

With PROC SQL it's even easier. All that's needed is a single target variable in the INTO clause, followed by the phrase SEPARATED BY and the separator specification in quotes. For example:

```
PROC SQL;
SELECT      fname
INTO        : fnames SEPARATED BY ' / '
FROM        thirteens
;
QUIT;


%PUT FNAMES: [&fnames];
```

This code also produces:

```
FNAMES:      [Alice / Barbara / Jeffrey]
```

DICTIONARY Tables

PROC SQL makes available a series of data sources that provide access to a great deal of metadata (that is, data **about** such system and data entities as libraries, tables, and columns). The existence and utility of these metadata are both completely independent of the macro facility. However, we are going to look at the metadata briefly at this point because it turns out that their value can often be leveraged significantly by passing their content into macro variables.

> **Reference:** The DICTIONARY tables are enumerated and described in the *Base SAS 9.2 Procedures Guide*: The SQL Procedure: Concepts: SQL Procedure: Using the DICTIONARY Tables.

We can illustrate the use of this metadata resource by reconsidering the initial example in this chapter. Recall that we had a table (named WIDE) with multiple numeric variables, and we wanted to generate a SELECT list to sum each of them without entering a lot of repetitious code. We devised a macro to do the job, but it depended on the columns following a naming pattern with numerically sequential suffixes. Now we deliberately undermine that solution by renaming the columns. We run:

```
PROC DATASETS;
MODIFY wide;
RENAME Measure1 = Estimated
       Measure2 = Net
       Measure3 = Gross
       Measure4 = Adjusted
       ;
RUN;
QUIT;
```

As a result, our table now looks like Exhibit 8-4.

Exhibit 8-4 WIDE

ID	Estimated	Net	Gross	Adjusted
A	11	12	13	14
B	21	22	23	24

The names of the numeric columns no longer form a series with successive integer suffixes, so the macro we developed earlier no longer serves to help SQL sum the columns of the table.

We have to generate code referring to individual columns one at a time, so DICTIONARY.COLUMNS is the appropriate source of metadata for this task. However, it contains such metadata for many columns, including columns in tables other than WIDE (or, possibly, in tables named WIDE that reside in libraries other than WORK). Fortunately, using SQL, we can apply a WHERE clause to restrict the yield. If we run:

```
PROC SQL;
SELECT        name, type
FROM          DICTIONARY.COLUMNS
WHERE         libname  = 'WORK' AND
              memname  = 'WIDE'
;
QUIT;
```

we see:

```
                                     Column
Column Name                          Type
------------------------------------------------
ID                                   char
Estimated                            num
Net                                  num
Gross                                num
Adjusted                             num
```

In particular, we see the names of the four columns of interest. We also see that the WHERE clause must be made a bit more restrictive to exclude character columns. So let's make that change, and at the same time include an INTO clause and the expression that will build the SELECT items we will ultimately need. So we now have:

```
PROC SQL;
SELECT        'sum(' || TRIM(name) || ') as Sum_' || name
INTO          : selections SEPARATED BY ', '
FROM          DICTIONARY.COLUMNS
WHERE         libname  = 'WORK' AND
              memname  = 'WIDE' AND
              type     = 'num'
;
QUIT;
```

It's pretty intricate, but less so if it's considered piece by piece. The expression following SELECT is like a template. It starts with a string constant:

```
sum(
```

and then drops in a column name (for example, "Gross") yielding:

```
sum(Gross
```

This is followed by another string constant, giving:

```
sum(Gross) as Sum_
```

and finally the repeated column name, so that the whole string is:

```
sum(Gross) as Sum_Gross
```

The INTO clause is straightforward. The WHERE clause restricts the query to the numeric columns of interest in the table of interest. After running the code, we can submit:

```
%PUT &selections;
```

In the log, this displays:

```
sum(Estimated) as Sum_Estimated,
sum(Net) as Sum_Net,
sum(Gross) as Sum_Gross,
sum(Adjusted) as Sum_Adjusted
```

So our query to process our collection of arbitrarily named columns can simply be:

```
PROC SQL;
CREATE TABLE sums AS
SELECT       &selections
FROM         wide
;
QUIT;
```

Running it yields the table SUMS as shown in Exhibit 8-5.

Exhibit 8-5 SUMS

Sum_Estimated	Sum_Net	Sum_Gross	Sum_Adjusted
32	34	36	38

One SQL statement has tapped the DICTIONARY COLUMNS table and used it to generate part of a subsequent SQL statement.

Automatic Macro Variables

Up to this point, we've been working with user-defined macro variables. PROC SQL also populates a number of automatically created macro variables. Let's look at a few of these (others relate specifically to the Pass-Through Facility, which is beyond the scope of this book).

Reference: Details about these automatic macro variables can be found in the *Base SAS 9.2 Procedures Guide*: Procedures: The SQL Procedure: Concepts: SQL Procedure: Using Macro Variables Set by PROC SQL.

The automatic macro variable &SQLRC echoes the return code set by the most recently processed SQL statement. Return codes are widely used in data processing to provide terse, digital status reports. Typically, a return code of zero is used to report success and the absence of problems, and PROC SQL conforms to this convention. To illustrate, we can run this code:

```
PROC SQL;
SELECT       MAX(height)
FROM         thirteens
;
QUIT;
```

There is no problem of any kind with this query, so if we then submit:

```
%PUT SQL Return Code is &sqlrc;
```

we see, in the log:

```
SQL Return Code is 0
```

Now let's see what happens when we intentionally create some difficulties. Consider this code:

```
PROC SQL;
CREATE TABLE maxmin AS
SELECT       MAX(height) AS m_height,
             MIN(height) AS m_height
FROM         thirteens
;
QUIT;
```

We've given the same name (M_HEIGHT) to two columns. This is acceptable within a SELECT statement, but it causes a problem when results are stored via a CREATE TABLE statement. Indeed, the log reports:

```
WARNING: Variable m_height already exists on file
         WORK.MAXMIN.
NOTE: Table WORK. MAXMIN created, with 1 rows and 1
      columns.
```

If we run the previous %PUT statement now, we see:

```
SQL Return Code is 4
```

In other words, 4 is the return code for a WARNING. Next, let's see what happens when we trigger an ERROR by asking the UNION operator to align a numeric column with a character column:

```
PROC SQL;
SELECT      fname
FROM        thirteens
UNION
SELECT      height
FROM        thirteens
;
QUIT;
```

We see, in the log:

```
ERROR: Column 1 from the first contributor of UNION is
not the same type as its counterpart from the second.
```

and our %PUT statement produces:

```
SQL Return Code is 8
```

So 8 apparently is the return code for an ERROR. Actually, it's a bit more complicated. Although the code in this example is syntactically correct, PROC SQL is nevertheless able to detect the problem as soon as it inspects the columns NAME and AGE. It does not have to look at any data values (names and ages recorded in the rows of THIRTEENS) to know that something is amiss; it has to see only the data types and other characteristics (that is, metadata).

In contrast, other types of errors become apparent only during actual processing of data. Consider this query:

```
PROC SQL;
SELECT      *,
            (SELECT fname
             FROM   thirteens
             WHERE  height>60)
FROM        thirteens
;
QUIT;
```

This is perfectly acceptable if the subquery within the parentheses returns a single row. In fact, the subset of THIRTEENS with height values over 60 consists of multiple rows. As a consequence, the query is not valid. However, the SQL processor does not know that until it opens the table and starts reading HEIGHT values. So the log reports:

```
ERROR: Subquery evaluated to more than one row.
```

Then, if we submit the %PUT statement, we see:

```
SQL Return Code is 16
```

Thus, we see that such run-time errors have their own return code of 16.

A return code of 24 differentiates yet another kind of error: one encountered by the host system. Those are a little harder to induce in a reproducible way, so we do not present an example here. Just suppose that we had a query that formed a huge intermediate result; that's easy enough to do when forming Cartesian products of even moderately sized tables. The SQL processor uses temporary disk storage to hold such intermediate results. If the intermediate results were to exhaust the available disk space, a return code of 24 would arise.

Two other return codes (12 and 28) are possible but rare. They indicate unanticipated internal error conditions—in other words, bugs in PROC SQL. They should be reported to SAS Technical Support.

Another automatic macro variable created by PROC SQL is &SQLOBS, which reports the number of rows processed by the most recent SQL statement. Consider this code:

```
PROC SQL;
SELECT      student.fname,
            classmate.fname
FROM        thirteens AS student
            JOIN
            thirteens AS classmate
ON          student.fname NE classmate.fname
WHERE       student.height>60 and classmate.weight<90
;
QUIT;
```

It returns three rows, so running:

```
%PUT macro variable SQLOBS: [&sqlobs];
```

displays:

```
macro variable SQLOBS: [3]
```

The last SQL automatic macro variable we consider here, &SQLOOPS, is a bit more obscure. It reports on "the number of iterations that the inner loop of PROC SQL executes." That depends not only on what the query is asking and on the volume of data it must process, but also on the methods used by the SQL processor in the background to actually perform the work. In other words, &SQLOOPS involves internal details and performance considerations. Those subjects are largely beyond the scope of this book.

8.3 Summary

The SAS macro facility works with PROC SQL in some very useful ways. Of course, macros can be used to generate PROC SQL code. SQL code can be used to populate macro variables, both singly and in groups and series. That capability provides a very useful way to use the metadata available via DICTIONARY tables. Automatic macro variables provide information on the outcome of PROC SQL statements.

Chapter 9

Table Maintenance and Alternate Strategies

It is often impossible to complete a SAS task in a single step. Typically, you need to make a series of incremental advances, each one introducing some refinement or extension. This raises an issue: how to materialize these increments in SAS data sets (tables).

SAS offers us three choices:

- succession
- replacement
- persistence

When we use a succession strategy, each table in the sequence is given a new and distinct identity. With replacement, a new table is given the same name as its predecessor, and it takes the place of that table. Persistence is altogether different in that there is no new table. Instead, changes are made to the existing table in place, without breaking the continuity of its existence.

This three-way distinction is an oversimplification. In practice, it's not unusual for a task to be accomplished using a mixture of two or even all three approaches. Nevertheless, it's important to appreciate the distinctions; it will help you to understand a number of PROC SQL features and their non-SQL counterparts.

To start with an extremely simple example, let's create a data set:

```
DATA simple;
DO Measure = 1 to 3; OUTPUT; END;
RUN;
```

Now suppose that the incremental refinement is to apply a label to the variable MEASURE. Using a **succession** strategy, we might run this DATA step to create a new data set with a distinct name (SIMPLE_BUT_LABELED):

```
DATA simple_but_labeled;
SET simple;
LABEL measure = 'Level reported after calibration';
RUN;
```

With a **replacement** strategy, we use the same code, except that the data set incorporating the label has the same name (SIMPLE) as the data set lacking the label:

```
DATA simple;
DO Measure = 1 to 3; OUTPUT; END;
RUN;

DATA simple;
SET  simple;
LABEL measure = 'Level reported after calibration';
RUN;
```

Things look quite different if we are using the **persistence** approach. We still create the initial data set the same way:

```
data simple;
do Measure = 1 to 3; output; end;
run;
```

However, to change the data set in place, we use PROC DATASETS to record the label:

```
PROC DATASETS;
MODIFY simple;
   LABEL measure = 'Level reported after calibration';
   RUN;
QUIT;
```

When we follow the persistence strategy and make changes to existing tables, we can say that we are conducting table **maintenance**. The bulk of this chapter is devoted to demonstrating the SQL table maintenance capabilities and their non-SQL counterparts.

The changes we can make to tables can be grouped into four categories:

- data changes
- metadata changes
- structural changes
- feature additions and removals

Data changes involve inserting new rows, deleting existing rows, or revising values within existing rows. Metadata changes are essentially modifications to the column attributes stored in table headers; our example using PROC DATASETS to record a label performed a metadata change. Structural changes typically involve removal of existing columns and creation of additional columns. The term "feature" refers to indexes, integrity constraints, and audit trails; a related topic is the use of generation data sets.

Use of the persistence strategy, and consequent reliance on table maintenance capabilities, is actually not that common. If you look at SAS code presented in papers at user conferences or in online postings, you are much more likely to see tables that are created and populated at the same time and never changed subsequently (in other words, reliance on the succession and replacement strategies). In part, that's probably because early versions of SAS provided no table maintenance tools other than the ability to change metadata via PROC DATASETS. In contrast, SQL "grew up" in the RDBMS (relational database management system) world, where the persistence approach is commonplace. Now SQL is part of SAS, and SAS offers table maintenance tools, whether you are using PROC SQL or other parts of SAS.

Before a table can be maintained in any way, it has to exist. When the persistence strategy is in use, it makes sense to think of the creation of a table as an event unto itself. Populating the table can then be treated as a subsequent maintenance task. In some situations this distinction is a very necessary one because there is a division of responsibility between administrators and maintenance programmers. The programmers might not have permission to create (or destroy) tables located in shared file spaces. So we devote a portion of this chapter to the life cycle of a table.

Keep in mind, however, that persistence does not necessarily imply **permanence**. The persistence strategy can be in effect for a table that is located in the WORK library and that exists for only a few seconds. On the other hand, a table kept in a permanent library for months or years might be a product of the succession or replacement strategies.

Finally, let's understand that the purpose of this chapter is not to advocate in general for the persistence strategy. Each of the three alternatives (succession, replacement, and persistence) has its advantages and disadvantages; the choice depends on the circumstances, and often a mixed approach is best. Our goal is just to look at the maintenance tools that support the persistence strategy.

9.1 Environment for Examples

Most, if not all, of the examples in previous chapters used the default WORK library, primarily for the sake of simplicity. In this chapter, we switch to a permanent library. In part, that's for the sake of variety. But the main reason for using a permanent library is to make the system option REPLACE | NOREPLACE operative (it has no effect on the WORK library).

So the first task is to create the library. On a Windows system, something like this should work:

```
LIBNAME demolib "c:\temp\demolib";
```

On other host systems, some adaptation is needed.

Because we are making changes to tables, we frequently want to refresh things, so that the effects of one demonstration don't contaminate the environment for subsequent demonstrations. For convenience, we can load the needed code into a simple macro:

```
%MACRO refresh_example;

PROC DATASETS LIBRARY=demolib KILL;
RUN;
```

```
PROC SQL;
CREATE TABLE demolib.fifteenups AS
SELECT      name AS FName, sex, age, height, weight
FROM        sashelp.class
WHERE       age GE 15
;
QUIT;


%MEND refresh_example;
```

The first step empties our **DEMOLIB** library. The second step creates a small table named FIFTEENUPS, which is the target for nearly all of our maintenance attempts. It looks like Exhibit 9-1.

Exhibit 9-1 DEMOLIB.FIFTEENUPS

FName	Sex	Age	Height	Weight
Janet	F	15	62.5	112.5
Mary	F	15	66.5	112.0
Philip	M	16	72.0	150.0
Ronald	M	15	67.0	133.0
William	M	15	66.5	112.0

Unless there is indication to the contrary, assume that this macro has been invoked immediately before each example in this chapter that involves the table FIFTEENUPS.

We also need a few other tables to drive some of the maintenance operations:

```
DATA insertions;
INFORMAT sex $1.;
INPUT FName $ Sex Height Weight Age;
CARDS;
Adam M 68 130 15
Joan F 64 120 16
;

DATA deletions;
INPUT FName $;
CARDS;
Mary
Ronald
;
```

```
DATA corrections;
INPUT FName $ Height Weight;
CARDS;
Janet    64   .
William  . 118
    ;
```

9.2 Distinguishing Persistence from Replacement

Many SAS users look at code performing replacement and perceive it as doing maintenance on an existing table. We can demonstrate the contrary by attempting a simple metadata declaration in the form of a FORMAT statement:

```
OPTIONS NOREPLACE;
DATA demolib.fifteenups;
SET  demolib.fifteenups;
FORMAT height weight 3.;
RUN;
```

In the log we see this:

```
WARNING: Data set DEMOLIB.FIFTEENUPS was not replaced
because of NOREPLACE option.
```

If we run a PROC PRINT on the table, we get this:

```
Obs    FName    Sex    Age    Height    Weight

 1     Janet     F     15     62.5      112.5
 2     Mary      F     15     66.5      112.0
 3     Philip    M     16     72.0      150.0
 4     Ronald    M     15     67.0      133.0
 5     William   M     15     66.5      112.0
```

Clearly the new formats are not in place. The DATA step was trying to create a new data set named DEMOLIB.FIFTEENUPS to replace the existing data set of the same name. Because the system option NOREPLACE was in effect, this failed.

We can attempt the same task in a way that avoids replacing the data set and instead adheres to the persistence strategy:

```
OPTIONS NOREPLACE;
PROC DATASETS LIBRARY=demolib;
MODIFY fifteenups;
   FORMAT height weight 3.;
   RUN;
QUIT;
```

Now the log tells us:

```
NOTE: MODIFY was successful for DEMOLIB.FIFTEENUPS.DATA.
```

and PROC PRINT displays this:

```
Obs    FName     Sex    Age    Height    Weight

1      Janet     F      15     63        113
2      Mary      F      15     67        112
3      Philip    M      16     72        150
4      Ronald    M      15     67        133
5      William   M      15     67        112
```

Notice that the new formats are in effect.

So, when data set replacement is not permitted, the persistence strategy (that is, maintaining an existing table in place) can succeed while the replacement strategy fails. We've modeled that using the NOREPLACE system option, but the situation (inability to replace a table in its entirety) can also arise from policies implemented by administrators using RDBMS or host operating system tools.

Even when replacement is possible (the situation we could bring about by toggling the system option to a value of REPLACE), in-place maintenance can be advantageous. The DATA step we tried to use to establish our new formats has to transcribe every observation from the existing DEMOLIB.FIFTEENUPS data set to the new one. That's inconsequential with five observations, but were there five million observations we'd notice the time needed.

While we're talking about the replacement strategy, let's briefly explore its use in PROC SQL. Here is code that replaces a table with a copy of itself:

```
OPTIONS REPLACE;
PROC SQL;
CREATE TABLE demolib.fifteenups AS
SELECT       *
FROM         demolib.fifteenups;
QUIT;
```

The log informs us:

```
WARNING: This CREATE TABLE statement recursively
references the target table. A consequence of this is a
possible data integrity problem.
```

but nevertheless indicates success:

```
NOTE: Table DEMOLIB.FIFTEENUPS created, with 5 rows and
5 columns.
```

The code is syntactically valid, and appears to work, but the warning message is sufficiently disturbing to make most users avoid such table-replacement code. That's why you won't see such recursive usage of the CREATE TABLE statement in this book.

Preview: We revisit the subject of SQL table-replacement code when we discuss generation data sets later in this chapter (see Section 9.7).

9.3 Life Cycle of a Table

When the persistence strategy is used, creating a table and populating it with rows of data can be distinct events. A complete life cycle also includes removal of the table.

Creation Using a Query

In earlier chapters we have in many situations created and populated new tables using the CREATE TABLE statement with the AS keyword. To illustrate, we can refresh our DEMOLIB library and then run this SQL statement:

```
PROC SQL;
CREATE TABLE not_empty AS
SELECT      SUBSTR(fname,1,1) LENGTH=1 AS Initial,
            sex,
            height
FROM        demolib.fifteenups
WHERE       age=15
;
QUIT;
```

The log tells us:

```
NOTE: Table WORK.NOT_EMPTY created, with 4 rows and 3
columns.
```

Because there are four 15-year-old students, we get four rows in the result. However, there is nothing about the code that ensures that the table will contain data. The oldest student is 16, so if we change the code to look for AGE values greater than that:

```
PROC SQL;
CREATE TABLE empty AS
SELECT      SUBSTR(fname,1,1) LENGTH=1 AS Initial,
            sex,
            height
FROM        demolib.fifteenups
WHERE       age=17
;
QUIT;
```

we get:

```
NOTE: Table WORK.EMPTY created, with 0 rows and 3
columns.
```

Now suppose that our purpose is to create, for later use, a table that we know to be empty. We can use something like this:

```
PROC SQL;
CREATE TABLE empty AS
SELECT      '' LENGTH=1 AS Initial,
            sex,
            height
FROM        demolib.fifteenups
WHERE       0
;
QUIT;
```

Notice that we have replaced the SUBSTR function reference with a null string to define the content of the new INITIAL column (because there are no rows stored, it doesn't matter, so we can keep it simple). The more important change is the placement of a simple 0 (zero) as the predicate of the WHERE clause; 0 (zero) is equivalent to FALSE when considered as a logical value, so this ensures that no rows are passed into the new table.

This form of the CREATE TABLE statement (with AS) is not intended for this purpose, but it turns out to be a really flexible way to generate new empty tables. In this example, we were able to establish one new column (INITIAL) from scratch, while modeling two others (SEX and HEIGHT) on columns in an existing table. We'll see that this is essentially a middle ground between two other forms of the CREATE TABLE statement.

Before turning to those techniques, let's look at the DATA step equivalent to the SQL statement we just used to create an empty table with a mix of old and new columns:

```
DATA empty;
STOP;
LENGTH Initial $ 1;
SET demolib.fifteenups;
KEEP initial sex height;
RUN;
```

This is somewhat odd-looking code, but in fact it works. We see in the log:

```
NOTE: The data set WORK.EMPTY has 0 observations and 3
variables.
```

Creation Using a Model

PROC SQL also provides a simple way to generate an empty table that is strictly based on an existing table. We simply replace AS with LIKE, followed by the name of the model. For example, if we run:

```
PROC SQL;
CREATE TABLE fifteenups_clone
LIKE        demolib.fifteenups
;
QUIT;
```

we get an empty copy of our model table (in other words, a table with none of the rows from the source table, but with all of the columns). The log confirms this:

```
NOTE: Table WORK.FIFTEENUPS_CLONE created, with 0 rows
and 5 columns.
```

We already know how to do this in a DATA step, just by simplifying the code we developed earlier:

```
DATA fifteenups_clone;
STOP;
SET demolib.fifteenups;
RUN;
```

This note appears in the log:

```
NOTE: The data set WORK.FIFTEENUPS_CLONE has 0
observations and 5 variables.
```

Creation Using Specifications

The third and final method for creating a new, empty table works entirely from coded specifications. In other words, it does not reference an existing table. Here is a simple example:

```
PROC SQL;
CREATE TABLE from_scratch
            (
               First   CHARACTER(10)
                       LABEL='Label for 1st column',
               Second NUMERIC
                       FORMAT=7.2
            )
   ;
   QUIT;
```

The log reports success:

```
NOTE: Table WORK.FROM_SCRATCH created, with 0 rows and 2
columns.
```

We have a character column named FIRST and a numeric column named SECOND. Naming and typing of the columns is mandatory. Specification of other attributes (LABEL, FORMAT, INFORMAT) is optional.

The equivalent DATA step is:

```
DATA from_scratch;
ATTRIB First  LENGTH = $ 10
               LABEL='Label for 1st column'
       Second FORMAT=7.2;
STOP;
RUN;
```

Help from the DESCRIBE TABLE Statement

PROC SQL provides a helpful (though potentially dangerous) tool in the form of the DESCRIBE TABLE statement. It examines an existing table, then displays (in the log) the CREATE TABLE statement that would create that table in an empty state (that is, with no rows of data). For example, if we run:

```
PROC SQL;
DESCRIBE TABLE demolib.fifteenups;
QUIT;
```

we get, in the log:

```
create table DEMOLIB.FIFTEENUPS
  (
   FName char(8),
   Sex char(1),
   Age num,
   Height num,
   Weight num
  );
```

You can copy this code from the log and use it in your source code. Of course, if you do that without making any changes, and if you do not have the NOREPLACE system option or some other safety mechanism in effect, you can easily destroy the model table (the one referenced by the DESCRIBE TABLE statement) and replace it with an empty clone. So it's more likely that you will want to edit the code before running it and, in particular, change the name or library location of the new table.

The DESCRIBE TABLE statement is useful in a variety of situations. If you want to code a self-contained CREATE TABLE statement for a table that is to be almost but not exactly like an existing table, you can use DESCRIBE TABLE to generate a first approximation of your CREATE TABLE statement. You can then edit the statement. If you want to pattern a new table on an existing table and then "freeze" that borrowed structure so that your code is insulated from later changes to the model table, you cannot use the CREATE TABLE statement with the LIKE keyword. Instead, use DESCRIBE TABLE to generate the code you need. If you prefer to design a table using the DATA step and then translate the specifications into SQL, use DESCRIBE TABLE to produce the CREATE TABLE statement.

Preview: The DATA step has an unrelated DESCRIBE statement. It is discussed later, in the chapter about views (see Section 10.3).

Deleting a Table

Before we proceed to an exploration of maintenance events, let's finish the discussion of the table life cycle by showing how a table can be eliminated when no longer needed.

The non-SQL technique relies on PROC DATASETS. To illustrate, we can run:

```
PROC DATASETS LIBRARY=demolib;
DELETE fifteenups;
RUN;
```

The log confirms the operation, reporting:

```
NOTE: Deleting DEMOLIB.FIFTEENUPS (memtype=DATA).
```

The PROC SQL counterpart is:

```
PROC SQL;
DROP TABLE    demolib.fifteenups
;
QUIT;
```

and the report in the log is:

```
NOTE: Table DEMOLIB.FIFTEENUPS has been dropped.
```

9.4 Data Maintenance

We're now ready to look at the core subject of this chapter, the SQL tools for changing the data content of a table (that is, the numeric or character values stored in rows and columns). There are three task types:

- inserting additional rows
- deleting existing rows
- making value changes within existing rows

SQL has a different statement for each task.

To simplify and shorten our discussion, we present only what might be called "nice" problems. That means we avoid having non-distinct keys (that is, keys that repeat) and we do not try to delete or change rows that do not exist, or to reference columns that do not exist. Keep in mind that in messier situations, some of the equivalences between SQL and DATA step code do break down. In real-world usage, you would of course determine the properties of your data and test the behavior of your code in the context of all potential ambiguities or anomalies.

Inserting Rows into an Existing Table

There are several ways to add completely new rows to a table. One of these is to specify the values for the new rows within the code. To demonstrate this, we first refresh our demonstration table (DEMOLIB.FIFTEENUPS) so that it looks like Exhibit 9-1.

Now suppose we have two new students, Adam and Joan. We can add their data to this table by running:

```
PROC SQL;
INSERT INTO  demolib.fifteenups
SET          fname   = "Adam"  ,
             weight  = 130     ,
             sex     = "M"     ,
             height  = 68      ,
             age     = 15
SET          fname   = "Joan"  ,
             sex     = "F"     ,
             age     = 16      ,
             height  = 64      ,
             weight  = 120
;
QUIT;
```

Notice that this is all one statement, but that there is a SET clause for each row to be inserted. The log confirms the operation:

```
NOTE: 2 rows were inserted into DEMOLIB.FIFTEENUPS.
```

DEMOLIB.FIFTEENUPS now looks like Exhibit 9-2.

Exhibit 9-2 DEMOLIB.FIFTEENUPS (after insertions)

FName	Sex	Age	Height	Weight
Janet	F	15	62.5	112.5
Mary	F	15	66.5	112.0
Philip	M	16	72.0	150.0
Ronald	M	15	67.0	133.0
William	M	15	66.5	112.0
Adam	M	15	68.0	130.0
Joan	F	16	64.0	120.0

The same thing can be done in a **DATA** step. To demonstrate, we first refresh the target (so that it has only its original five rows), and then run:

```
DATA demolib.fifteenups;
fname    = "Adam"  ;
sex      = "M"     ;
age      = 15      ;
height   = 68      ;
weight   = 130     ;
OUTPUT;
fname    = "Joan"  ;
sex      = "F"     ;
age      = 16      ;
height   = 64      ;
weight   = 120     ;
OUTPUT;
STOP;
MODIFY demolib.fifteenups;
RUN;
```

The log reports:

```
NOTE: The data set DEMOLIB.FIFTEENUPS has been updated.
There were 0 observations rewritten, 2 observations
added and 0 observations deleted.
```

The result is the same seven-row table we saw produced by SQL (see Exhibit 9-2).

There is also a more compact way of coding SQL to perform this task: the VALUES clause. The code looks like this:

```
INSERT INTO   demolib.fifteenups
VALUES        ("Adam", "M", 15, 68, 130)
VALUES        ("Joan", "F", 16, 64, 120)
;
```

As before, the log tells us:

```
NOTE: 2 rows were inserted into DEMOLIB.FIFTEENUPS.
```

Notice that in this terse code there is no indication of which column is to receive which value. It's obvious to us that "Adam" is a name, and that 16 is too small to be either a height or a weight, and thus must be an age. The SQL processor is not that smart. It just loads the values into columns in the order in which the columns are stored. We managed to specify the values in the appropriate order, but relying on that is rather risky. It's better to extend the INTO clause with a column list (which is optional syntax). The code then looks like this:

```
INSERT INTO   demolib.fifteenups
              (fname, sex, age, height, weight)
VALUES        ("Adam", "M", 15, 68, 130)
VALUES        ("Joan", "F", 16, 64, 120)
;
```

With this explicit technique, we don't have to be concerned with the pre-existing order of the columns in the target table. As long as the order of the column names in the INTO clause corresponds to the order of the literals in the VALUES clauses, the table will be extended correctly.

There is no DATA step counterpart to the VALUES clause.

All of the methods we've used so far to extend our table have embedded the new data directly into the programming statements. Such code-driven approaches might suffice when there is a one-time need to introduce a small volume of data. However, a more systematic approach calls for the new data to be introduced via a query (that is, using a SELECT clause to supply the values). We conclude our discussion about extending tables by demonstrating such data-driven techniques.

Remember that we earlier built a table (INSERTIONS) containing the data for Adam and Joan. It looks like Exhibit 9-3.

Exhibit 9-3 INSERTIONS

Sex	FName	Height	Weight	Age
M	Adam	68	130	15
F	Joan	64	120	16

After refreshing the target table (DEMOLIB.FIFTEENUPS) to its initial state, we can try to integrate the data for the two new students by running:

```
PROC SQL;
INSERT INTO  demolib.fifteenups
SELECT       *
FROM         insertions
;
QUIT;
```

The log shows:

```
NOTE: 2 rows were inserted into DEMOLIB.FIFTEENUPS.
```

The log gives no indication of trouble, but DEMOLIB.FIFTEENUPS looks like Exhibit 9-4.

Exhibit 9-4 DEMOLIB.FIFTEENUPS (with incorrect insertions)

FName	Sex	Age	Height	Weight
Janet	F	15	62.5	112.5
Mary	F	15	66.5	112.0
Philip	M	16	72.0	150.0
Ronald	M	15	67.0	133.0
William	M	15	66.5	112.0
M	A	68	130.0	15.0
F	J	64	120.0	16.0

The problem is that the columns in the two tables are not stored in the same order and if there is no guidance to the contrary, SQL matches them up by position, not by name. Fortunately, it is possible to avoid this problem by coding an explicit column list within the INTO clause, and we already know that the SELECT clause can be made similarly explicit. So let's try this:

```
PROC SQL;
INSERT INTO  demolib.fifteenups
             (age, height, fname, sex, weight)
SELECT        age, height, fname, sex, weight
FROM          insertions
;
QUIT;
```

We get the same note, but this time the extended table is correct, and again looks like Exhibit 9-2.

SAS also offers a non-SQL tool for reading observations from one data set and inserting them into an existing data set: PROC APPEND. We can test it (after refreshing the target table to its original state) by running:

```
PROC APPEND BASE=demolib.fifteenups DATA=insertions;
RUN;
```

The log tells us:

```
NOTE: Appending WORK.INSERTIONS to DEMOLIB.FIFTEENUPS.
NOTE: There were 2 observations read from the data set
      WORK.INSERTIONS.
NOTE: 2 observations added.
```

The differing order of the variables causes no problem. Unlike PROC SQL, PROC APPEND automatically relies on names to match up the variables.

Deleting Rows

Suppose that instead of inserting new rows, we find it necessary to delete some of the existing rows. As with insertion, we find that there are both code-driven and data-driven techniques.

Let's say that we have to delete the rows for Mary and Ronald from the DEMOLIB.FIFTEENUPS table. We can run this DATA step:

```
DATA demolib.fifteenups;
MODIFY demolib.fifteenups;
WHERE fname IN ('Mary','Ronald');
REMOVE;
RUN;
```

The log reports:

```
NOTE: There were 2 observations read from the data set
      DEMOLIB.FIFTEENUPS.
      WHERE fname in ('Mary', 'Ronald');
NOTE: The data set DEMOLIB.FIFTEENUPS has been updated.
      There were 0 observations rewritten, 0
      observations added and 2 observations deleted.
```

The table now looks like Exhibit 9-5.

Exhibit 9-5 DEMOLIB.FIFTEENUPS (after deletions)

FName	Sex	Age	Height	Weight
Janet	F	15	62.5	112.5
Philip	M	16	72.0	150.0
William	M	15	66.5	112.0

The equivalent SQL code (which we run after refreshing DEMOLIB.FIFTEENUPS) is:

```
PROC SQL;
DELETE FROM  demolib.fifteenups
WHERE        fname IN ('Mary','Ronald')
;
QUIT;
```

It generates this log note:

```
NOTE: 2 rows were deleted from DEMOLIB.FIFTEENUPS.
```

The deletion process can also be data-driven. This requires a table containing the keys (in this instance, FNAME values) of the rows to be deleted. We built such a table (DELETIONS) earlier. It looks like Exhibit 9-6.

Exhibit 9-6 DELETIONS

FName
Mary
Ronald

To demonstrate how this table can drive the deletion process in a DATA step, we first (of course) refresh the target table to its initial five-row population, and then run this code:

```
DATA demolib.fifteenups;
MODIFY demolib.fifteenups deletions(IN=dropit);
BY fname;
IF dropit THEN REMOVE;
RUN;
```

In the log we see:

```
NOTE: There were 1 observations read from the data set
      DEMOLIB.FIFTEENUPS.
NOTE: The data set DEMOLIB.FIFTEENUPS has been updated.
      There were 0 observations rewritten, 0
      observations added and 2 observations deleted.
NOTE: There were 2 observations read from the data set
      WORK.DELETIONS.
```

The first **NOTE** is totally misleading. For some reason, when SAS uses this form of the MODIFY statement, it seems to always report reading just one observation from the master data set. The important things to notice are that the second note conforms to our expectations and that the table ends up looking like Exhibit 9-5.

Now (after we refresh the target table) we want to see how the deletion can be done in PROC SQL. The code is:

```
PROC SQL;
DELETE FROM  demolib.fifteenups
WHERE        fname IN (SELECT fname FROM deletions)
;
QUIT;
```

When we run it, the log reports:

```
NOTE: 2 rows were deleted from DEMOLIB.FIFTEENUPS.
```

Selective Corrections

We've looked at insertions and deletions. Now we turn to the third type of change: replacement of existing values. Whereas insertion and replacement are done on a row basis, replacement can be done selectively **within** a row.

Let's take up a concrete example. Suppose that we have the (suitably refreshed) DEMOLIB.FIFTEENUPS table we've used before, and that we have two new pieces of data: Janet's height is now 64 inches and William's weight is now 118 pounds. Here's the code-driven DATA step approach:

```
DATA demolib.fifteenups;
MODIFY demolib.fifteenups;
IF fname='Janet' THEN DO;
   height = 64;
   REPLACE;
   END;
IF fname='William' THEN DO;
   weight = 118;
   REPLACE;
   END;
RUN;
```

In the log we see:

```
NOTE: There were 5 observations read from the data set
      DEMOLIB.FIFTEENUPS.
NOTE: The data set DEMOLIB.FIFTEENUPS has been updated.
      There were 2 observations rewritten, 0
      observations added and 0 observations deleted.
```

Turning to SQL, we can do the same thing with this code (after refreshing the target table):

```
PROC SQL;
UPDATE       demolib.fifteenups
SET          height =  64
WHERE        fname='Janet'
;
UPDATE       demolib.fifteenups
SET          weight = 118
WHERE        fname='William'
;
QUIT;
```

Because we need a separate statement for each row being changed, it's a little clumsy. In the log, each statement is followed by this note:

```
NOTE: 1 row was updated in DEMOLIB.FIFTEENUPS.
```

Whether we use the DATA step or PROC SQL, the table ends up looking like Exhibit 9-7.

Exhibit 9-7 DEMOLIB.FIFTEENUPS (after corrections)

FName	Sex	Age	Height	Weight
Janet	F	15	64.0	112.5
Mary	F	15	66.5	112.0
Philip	M	16	72.0	150.0
Ronald	M	15	67.0	133.0
William	M	15	66.5	118.0

Embedding value corrections in code might be a useful approach for a quick fix involving a small amount of data. However, a systematic approach for larger volumes calls for a process that is data-driven, with the corrections stored in a separate table. We already built such a table (CORRECTIONS) for our example. It is shown in Exhibit 9-8.

Exhibit 9-8 CORRECTIONS

FName	Height	Weight
Janet	64	.
William	.	118

Note the missing values for Janet's weight and William's height. They are just placeholders; the corresponding values in DEMOLIB.FIFTEENUPS should be left as is.

To accomplish this with a DATA step, we can refresh the target table and run:

```
DATA    demolib.fifteenups;
MODIFY demolib.fifteenups
       corrections;
BY fname;
RUN;
```

which places these notes in the log:

```
NOTE: There were 1 observations read from the data set
      DEMOLIB.FIFTEENUPS.
NOTE: The data set DEMOLIB.FIFTEENUPS has been updated.
      There were 2 observations rewritten, 0
      observations added and 0 observations deleted.
NOTE: There were 2 observations read from the data set
      WORK.CORRECTIONS.
```

Once again the first NOTE is incorrect. However, the table does in fact reflect both changes. It now looks like Exhibit 9-7.

Developing parallel SQL language is a bit tricky. We can use this code:

```
PROC SQL;
UPDATE          demolib.fifteenups AS main
SET             height = COALESCE( (SELECT sub.height
                                    FROM corrections AS sub
                                    WHERE main.fname=sub.fname),
                                   height),
                weight = COALESCE( (SELECT sub.weight
                                    FROM corrections AS sub
                                    WHERE main.fname=sub.fname),
                                   weight)
;
QUIT;
```

Notice that there is a correlated subquery for each column in the SET clause (that is, one for HEIGHT and one for WEIGHT). The COALESCE function is used to prevent nulls (missing values) from replacing non-nulls. When we run the code we see:

```
NOTE: 5 rows were updated in DEMOLIB.FIFTEENUPS.
```

This tells us that the query examined all of the rows in the target table, not just those that were to be changed. In a real-world problem, this could be quite inefficient. A remedy is to insert this WHERE clause after the SET clause:

```
WHERE fname IN (SELECT fname FROM corrections)
```

If we do that, the confirmation message is:

```
NOTE: 2 rows were updated in DEMOLIB.FIFTEENUPS.
```

This confirms that PROC SQL used the subquery in the WHERE clause to restrict processing to the rows for which there were changes.

9.5 Metadata Maintenance

At this point we've finished our exploration of data maintenance (that is, techniques for inserting, deleting, and changing the rows within a table). Now we turn to the tools that allow us to work on the metadata (column attributes). It's important to understand that metadata are stored in the header portion of a table, so that they can be touched without disturbing the body of the table (the row/column grid containing the data values).

The primary tool for metadata management is PROC DATASETS and, in particular, its MODIFY statement. This statement is not to be confused with either the MODIFY statement of the DATA step language or the MODIFY clause of the SQL ALTER TABLE statement, which is discussed later in this section.

As an example, suppose we want to enhance our DEMOLIB.FIFTEENUPS table by providing a label and a format for the HEIGHT variable. We can do that with this code:

```
PROC DATASETS LIBRARY=demolib;
MODIFY fifteenups;
   FORMAT height 6.2;
   LABEL height = 'Height in Inches';
   RUN;
QUIT;
```

When we run this PROC DATASETS code, we see this in the log:

```
NOTE: MODIFY was successful for DEMOLIB.FIFTEENUPS.DATA.
```

PROC SQL has an ALTER statement, which in turn provides a MODIFY clause for making metadata changes. So the SQL code to establish a format and a label for HEIGHT would be:

```
PROC SQL;
ALTER TABLE   demolib.fifteenups
MODIFY        height FORMAT = 6.2
                     LABEL = 'Height in Inches'
  ;
QUIT;
```

When we run the code, the log reports:

```
NOTE: Table DEMOLIB.FIFTEENUPS has been modified, with 5
columns.
```

Tip: A MODIFY statement group in PROC DATASETS can include a RENAME statement. We do not include that in the example because PROC SQL has no corresponding feature. If you are using SQL and you want to change a column name, you have to exit from PROC SQL and work with PROC DATASETS.

9.6 Changing Structure

In previous sections we have worked with techniques to change data (the rows in the body of a table, and their content) and metadata (attributes of a table's column, stored in the table header). Now we consider changes in structure, meaning the number of columns, their types (numeric or character), and their lengths (meaning the number of bytes of storage allocated for each value in a column). It is not possible to change a column's type, but the ALTER TABLE statement in PROC SQL does provide DROP and ADD clauses. DROP is used to remove existing columns, and ADD to insert NEW columns. In addition, the MODIFY clause, which we used earlier to change metadata, can also be used to change a column's length. So, for example, we could change our DEMOLIB.FIFTEENUPS table by extending the length of the FNAME column, shedding the AGE, HEIGHT, and WEIGHT columns, and inserting a DOB column (for date of birth). The code for this is:

```
PROC SQL;
ALTER TABLE   demolib.fifteenups
MODIFY        fname CHAR (12)
DROP          age, height, weight
ADD           DoB DATE LABEL='Date of Birth'
;
QUIT;
```

Notice the specification of DATE in the place that syntactically expects a declaration of data type. DATE is a valid type in most if not all SQL implementations, but SAS of course has only the numeric and character types. So, if you call for a DATE column, PROC SQL automatically makes it numeric, and also by default associates an appropriate informat and format.

The code works as promised, as long as the REPLACE system option is in effect. The log confirms the completion of the task:

```
NOTE: Table DEMOLIB.FIFTEENUPS has been modified, with 3
columns.
```

The wording of this note suggests that the process did not disrupt the table's continuity of existence. However, if we put the NOREPLACE system option into effect and rerun this code (after running the %REFRESH_EXAMPLE macro to restore

DEMOLIB.FIFTEENUPS to its initial state), we get a different outcome. This time the log tells us:

```
WARNING: Data set DEMOLIB.FIFTEENUPS was not replaced
         because of NOREPLACE option.
ERROR:   PROC SQL cannot modify the data set due to the
         reason(s) cited above.
```

This tells us that, behind the scenes, PROC SQL actually has to replace a table in order to make structural changes specified in DROP, ADD, and MODIFY clauses. A consequence is that if you do not have permission to delete and replace a table, you probably won't be able to use all of the features of the ALTER statement.

There are no counterparts in PROC DATASETS to the DROP and ADD clauses. So the non-SQL equivalent to our PROC SQL code is a DATA step that explicitly calls for replacement. The code looks like this:

```
OPTIONS REPLACE;
DATA demolib.fifteenups;
LENGTH FName $ 12;
SET  demolib.fifteenups(DROP = age height weight);
ATTRIB DoB LABEL = 'Date of Birth'
           FORMAT = date.
           INFORMAT = date.;
RUN;
```

Notice the FORMAT and INFORMAT specifications, which did not appear in the PROC SQL code, where we instead specified DATE as the data type. Whether we use the ALTER TABLE statement in PROC SQL or the DATA step to create it, the new DEMOLIB.FIFTEENUPS looks like Exhibit 9-9.

Exhibit 9-9 DEMOLIB.FIFTEENUPS (with structure changes)

FName	Sex	DoB
Janet	F	.
Mary	F	.
Philip	M	.
Ronald	M	.
William	M	.

The new DOB column is of course null-filled. Supplying values is a separate data maintenance exercise.

Here is a portion of PROC CONTENTS output for DEMOLIB.FIFTEENUPS, reflecting the changes just made:

```
Alphabetic List of Variables and Attributes

#  Variable  Type  Len  Format  Informat  Label

3  DoB       Num     8  DATE.   DATE.     Date of Birth
1  FName     Char   12
2  Sex       Char    1
```

So, although table structure can be changed, doing so inevitably involves replacing the table. Thus it does not really fall under what we have termed the persistence strategy. If you are working in an environment where file permissions are restricted, you might find that you cannot complete such tasks on your own.

9.7 Changing Features

We conclude our discussion of table maintenance by discussing what we call, for want of a better term, "features." To be specific, this term refers to

- indexes
- integrity constraints
- audit trails
- generation data sets

All four of these features can be used with PROC SQL, though the extent of support varies. All, except generation data sets, are particularly useful with the persistence strategy to manage progression in your data.

Indexes

SAS indexes serve two purposes. The first is to improve performance by speeding things up. The second is to support BY processing without requiring the data set to be sorted. Because SQL does not use BY processing, or for that matter any order-sensitive processing, indexes in PROC SQL are about performance alone.

Reference: Read more about indexes in *SAS 9.2 Language Reference: Concepts*: SAS Files Concepts: SAS Data Files: Understanding SAS Indexes.

Outside PROC SQL, we use **PROC DATASETS** to add an index to an existing data set. Suppose we want to index our demonstration data set by the variable AGE. The code would be:

```
PROC DATASETS LIBRARY=demolib;
MODIFY fifteenups;
   INDEX CREATE age;
   RUN;
QUIT;
```

When we run this, the log reports:

```
NOTE: Simple index age has been defined.
```

We can do the same thing using SQL. In that case the code looks like this:

```
PROC SQL;
CREATE INDEX age
ON            demolib.fifteenups
;
QUIT;
```

The confirmation message in the log is (again):

```
NOTE: Simple index age has been defined.
```

Let's confirm that the index actually works. If we run this code:

```
PROC PRINT DATA=demolib.fifteenups;
RUN;
```

to display the contents of DEMOLIB.FIFTEENUPS, we see:

Obs	FName	Sex	Age	Height	Weight
1	Janet	F	15	62.5	112.5
2	Mary	F	15	66.5	112.0
3	Philip	M	16	72.0	150.0
4	Ronald	M	15	67.0	133.0
5	William	M	15	66.5	112.0

The observations appear in FNAME alphabetical order, because they are stored in that order. If we include a BY statement in the code, as in:

```
PROC PRINT DATA=demolib.fifteenups;
BY age;
RUN;
```

the output looks like this:

```
Age=15

Obs     FName      Sex      Height      Weight

 1      Janet       F        62.5        112.5
 2      Mary        F        66.5        112.0
 4      Ronald      M        67.0        133.0
 5      William     M        66.5        112.0


Age=16

Obs     FName      Sex      Height      Weight

 3      Philip      M         72          150
```

We get this result because the index has been used to display the observations in ascending AGE order. Notice how the numbers in the "Obs" reference column reflect the stored order of the rows, not the displayed order.

Let's focus now on the use of indexes in PROC SQL (whether or not they are created with PROC SQL). As we noted earlier, the benefit of indexes in PROC SQL is strictly one of performance; indexes can speed up processing. Later (see Section 13.1), we have an example of this.

Often it's convenient to create, populate, and index a table all at once. In SAS generally, doing this simply involves using a data set option. To create an AGE index for a copy of DEMOLIB.FIFTEENUPS, we could run:

```
DATA newtable( INDEX=(age) );
SET demolib.fifteenups;
RUN;
```

SQL offers no syntax within the CREATE TABLE statement to call for construction of an index. However, because data set options are available within PROC SQL, we can borrow that SAS feature and use it in code like this:

```
PROC SQL;
CREATE TABLE newtable( INDEX=(age) ) AS
SELECT       *
FROM         demolib.fifteenups
;
QUIT;
```

Either way, NEWTABLE is created with an index on AGE.

There might be circumstances in which you want to get rid of an index while preserving the data set (and possibly other indexes). The non-SQL technique is to use the INDEX DELETE statement in PROC DATASETS. Here's a demonstration that creates two indexes and then eliminates one of them:

```
PROC DATASETS LIBRARY=demolib;
MODIFY fifteenups;
   INDEX CREATE age;
   INDEX CREATE sex;
   RUN;
MODIFY fifteenups;
   INDEX DELETE age;
   RUN;
QUIT;
```

We get these messages for the first MODIFY statement group:

```
NOTE: Simple index Age has been defined.
NOTE: Simple index Sex has been defined.
NOTE: MODIFY was successful for DEMOLIB.FIFTEENUPS.DATA.
```

Then, for the second group, we see:

```
NOTE: Index Age deleted.
NOTE: MODIFY was successful for DEMOLIB.FIFTEENUPS.DATA.
```

In PROC SQL, the DROP INDEX statement parallels the INDEX DELETE statement that we just saw in PROC DATASETS. This time we continue the example without refreshing. With the SEX index still in place, we have PROC SQL build and then remove the AGE index. The code is:

```
PROC SQL;
CREATE INDEX age
ON          demolib.fifteenups
;
DROP INDEX  age
FROM        demolib.fifteenups
;
QUIT;
```

After the first statement, the log reports:

```
NOTE: Simple index age has been defined.
```

Then, after the second statement, the note is:

```
NOTE: Index age has been dropped.
```

We conclude the discussion of indexes by looking at tools we can use to discover their presence. Once again, we demonstrate using DEMOLIB.FIFTEENUPS in its state at the conclusion of the previous example (that is, without running the refresh macro). Outside of PROC SQL, existence of indexes can be reported with PROC CONTENTS. If we run:

```
PROC CONTENTS DATA=demolib.fifteenups;
RUN;
```

the output includes this report:

```
Alphabetic List of Indexes and Attributes

                    # of
                  Unique
#     Index       Values

1     Sex            2
```

The corresponding tool in PROC SQL is the DESCRIBE TABLE statement. For example, we could submit:

```
PROC SQL;
DESCRIBE TABLE demolib.fifteenups;
QUIT;
```

Recall: We used this statement earlier in this chapter (see Section 9.3) in conjunction with creating tables modeled on existing tables.

The result, appearing in the log, begins with a complete CREATE TABLE statement outlining the structure of DEMOLIB.FIFTEENUPS, followed by:

```
create index Sex on DEMOLIB.FIFTEENUPS(Age);
```

which tells us that the index on AGE exists by showing us how it might have been created.

Integrity Constraints

Integrity constraints are devices that assure that data values in particular columns conform to various rules. For example, we can specify an integrity constraint to ensure that the values in a column are distinct (that is, that no value occurs more than once).

Integrity constraints cannot operate when a table is created and populated at the same time. They can be used **prospectively**, when data maintenance operations are performed on an existing table, or **retrospectively**, to check the compliance of the data stored in an existing table.

> **Reference:** Read more about integrity constraints in *SAS 9.2 Language Reference: Concepts*: SAS Files Concepts: SAS Data Files: Understanding Integrity Constraints.

We begin by demonstrating the life cycle of an integrity constraint, using non-SQL tools. After using our macro to refresh DEMOLIB.FIFTEENUPS, we run this code to create the constraint:

```
PROC DATASETS LIBRARY=demolib;
MODIFY fifteenups;
   IC CREATE norepeats = UNIQUE(fname);
   RUN;
QUIT;
```

The log reports:

```
NOTE: Integrity constraint norepeats defined.
```

There are several types of integrity constraints. The keyword UNIQUE indicates that here we want a constraint that prevents any observation in this data set from holding a value of the variable FNAME that repeats an FNAME value found in another observation. In other words, it ensures that each FNAME value is distinct. NOREPEATS is a name we supply so that we can later reference this integrity constraint. If we run:

```
PROC CONTENTS DATA=demolib.fifteenups;
RUN;
```

we see:

```
Alphabetic List of Integrity Constraints

     Integrity
#    Constraint    Type       Variables

1    norepeats     Unique     FName
```

Now let's see our integrity constraint in action. We attempt to insert an additional observation carrying the FNAME value "Ronald," which already appears in the table:

```
DATA demolib.fifteenups;
fname = 'Ronald';
OUTPUT;
STOP;
MODIFY demolib.fifteenups;
RUN;
```

In the log, we see:

```
FName=Ronald Sex=  Age=. Height=. Weight=. _ERROR_=1
_IORC_=660130 _N_=1

NOTE: The data set DEMOLIB.FIFTEENUPS has been updated.  There
were 0 observations rewritten, 0 observations added and 0
observations deleted.

NOTE: There were 0 rejected updates, 1 rejected adds, and 0
rejected deletes.
```

It is "1 rejected adds" which indicates that the integrity constraint blocked the OUTPUT statement from creating a duplicate observation.

Finally, if we want to remove the constraint, PROC DATASETS provides a tool to do so. The code looks like this:

```
PROC DATASETS LIBRARY=demolib;
MODIFY fifteenups;
   IC DELETE norepeats;
   RUN;
QUIT;
```

The log confirms the operation, reporting:

```
NOTE: All integrity constraints defined on
DEMOLIB.FIFTEENUPS.DATA have been deleted.

NOTE: MODIFY was successful for DEMOLIB.FIFTEENUPS.DATA.
```

Now we turn to PROC SQL, where we can use the same example. After refreshing the table, we first create the integrity constraint by submitting:

```
PROC SQL;
ALTER TABLE  demolib.fifteenups
ADD CONSTRAINT norepeats DISTINCT(fname)
;
QUIT;
```

The confirmation is rather vague in this case; the log states only:

```
NOTE: Table DEMOLIB.FIFTEENUPS has been modified, with 5
columns.
```

To verify that the integrity constraint has been put in place, we can use the DESCRIBE TABLE statement:

```
PROC SQL;
DESCRIBE TABLE CONSTRAINTS demolib.fifteenups;
QUIT;
```

Then, in the log, we see:

```
-----Alphabetic List of Integrity Constraints-----

     Integrity
#    Constraint    Type       Variables
-----------------------------------------
1    norepeats     Unique     FName
```

To demonstrate our integrity constraint in action, we can try:

```
PROC SQL;
INSERT INTO  demolib.fifteenups
SET          fname = "Ronald"
;
QUIT;
```

Because there is already a row for Ronald, the attempt to insert a second such row fails, and the log tells us:

```
ERROR: Add/Update failed for data set DEMOLIB.FIFTEENUPS
because data value(s) do not comply with integrity constraint
norepeats.

NOTE: Deleting the successful inserts before error noted above
to restore table to a consistent state.
```

The note is superfluous in this case, because there were no other inserts. It does tell us that, in general, the SQL processor treats insertions subject to integrity constraints as an all-or-nothing proposition.

To use SQL to remove the integrity constraint, we would run:

```
PROC SQL;
ALTER TABLE  demolib.fifteenups
DROP CONSTRAINT norepeats
;
QUIT;
```

Then the log shows us:

```
NOTE: Integrity constraint norepeats deleted.
```

That completes the SQL version of our life-cycle demonstration of an integrity constraint. However, there is one additional capability found in SQL that does not exist elsewhere in SAS. We can establish an integrity constraint when a table is first created (though not if either the LIKE or AS keyword is used in the CREATE TABLE statement). Here is an example:

```
PROC SQL;
CREATE TABLE AnotherTable
              ( SomeColumn character(20),
                CONSTRAINT norepeats UNIQUE(SomeColumn),
                AnotherColumn numeric
              )
;
QUIT;
```

Before leaving the subject of integrity constraints, let's look at their retrospective use. Such use is based on the fact that when an integrity constraint is created for a table that already contains data, SAS must confirm that the existing data are compliant with the integrity constraint. If the data are not compliant, creation of the integrity constraint is blocked. We can use this process to evaluate an existing table.

Let's experiment and demonstrate this technique. After refreshing DEMOLIB.FIFTEENUPS to its initial state, we can run:

```
PROC SQL;
ALTER TABLE  demolib.fifteenups
 ADD CONSTRAINT diff_age UNIQUE(age)
 ;
QUIT;
```

This requires that the AGE values in the table be distinct. However, we know that we have several 15-year-olds. So, in the log we see:

```
ERROR: Duplicate values not allowed on index Age for file
FIFTEENUPS.
```

Why does it refer to an index? Because integrity constraints are implemented through indexes.

If the AGE values had been distinct in the table, the operation to add the constraint would have succeeded, and we would not have gotten the ERROR message. So it's possible to assess a table's pre-existing compliance with an integrity constraint by attempting to install the constraint and monitoring the result of that attempt.

Audit Trails

A table's audit trail is an auxiliary data set that keeps track of the data changes made to the subject table. Audit trails are germane only when you are following the persistence strategy and performing data maintenance on an existing table in place. PROC SQL itself has no language elements to manage audit trails, but if a table has an audit trail, PROC SQL records in that audit trail changes it makes as it processes SQL INSERT, DELETE, and UPDATE statements.

> **Reference:** Read more about audit trails in *SAS 9.2 Language Reference: Concepts*: SAS Files Concepts: SAS Data Files: Understanding an Audit Trail.

To demonstrate, we can run this code (after refreshing DEMOLIB.FIFTEENUPS to its initial state):

```
PROC DATASETS LIBRARY=demolib;
AUDIT fifteenups;
   INITIATE;
   RUN;
QUIT;
```

This turns on the audit trail feature for DEMOLIB.FIFTEENUPS, as confirmed by this log message:

```
NOTE: The data set DEMOLIB.FIFTEENUPS.AUDIT has 0 observations
and 11 variables.
```

This empty data set is the container for the audit trail. Notice the three-part name. The third part, known as the Member Type, is "AUDIT," whereas the Member Type of our data table is "DATA." We have to use PROC DATASETS to initiate the audit trail; PROC SQL has no equivalent ability.

Now we can exercise our audit trail. We first use a DATA step to insert a new row in our data table. The code is:

```
DATA demolib.fifteenups;
fname = 'Ezra';
OUTPUT;
STOP;
MODIFY demolib.fifteenups;
RUN;
```

Now we display what's in both the data table and its audit trail. When we submit:

```
PROC PRINT DATA=demolib.fifteenups;
VAR fname;
RUN;
```

we see, as expected:

Obs	FName
1	Janet
2	Mary
3	Philip
4	Ronald
5	William
6	Ezra

Turning to the audit trail, we run:

```
PROC PRINT DATA=demolib.fifteenups(TYPE=AUDIT);
VAR fname _atdatetime_;
FORMAT _atdatetime_ tod.;
RUN;
```

This gives us:

Obs	FName	_ATDATETIME_
1	Ezra	15:29:50

There is just one row, corresponding to the only row in the audited table that changed after the audit trail was begun. The format statement directs PROC PRINT to display only the time-of-day portion of the timestamp, but the date portion is stored.

Now we test the use of the audit trail with PROC SQL. To make the audit trail cumulative, we do **not** refresh the data table. Instead, we proceed to insert two more rows by running:

```
PROC SQL;
INSERT INTO   demolib.fifteenups
SET           fname = 'Nicole'
SET           fname = 'Matthew'
;
QUIT;
```

We see confirmation in the log:

```
NOTE: 2 rows were inserted into DEMOLIB.FIFTEENUPS.
```

Now we rerun the two **PROC PRINT** steps. The first one operates on the data table; the output looks like:

```
Obs     FName

1       Janet
2       Mary
3       Philip
4       Ronald
5       William
6       Ezra
7       Nicole
8       Matthew
```

This output reflects the presence of the two new rows. The second displays the audit trail. We see:

```
Obs     FName       _ATDATETIME_

1       Ezra         15:29:50
2       Nicole       15:29:54
3       Matthew      15:29:54
```

The first observation remains from the earlier insertion (the one performed with a DATA step), whereas the other two report on the rows we just added using the INSERT statement in PROC SQL.

Now suppose we want to end the audit trail. PROC SQL cannot do that; we have to use PROC DATASETS. Here is the code:

```
PROC DATASETS LIBRARY=demolib;
AUDIT fifteenups;
    TERMINATE;
    RUN;
QUIT;
```

The log reports:

```
NOTE: Deleting DEMOLIB.FIFTEENUPS (memtype=AUDIT).
```

Notice that this does not merely "freeze" the audit trail. It actually deletes it. So if you need to preserve the content of a discontinued audit trail, you should copy it into a data set before you terminate it.

Generation Data Sets

To conclude this chapter examining tools for table maintenance, we look briefly at generation data sets. Actually, the topic of generation data sets does not belong here. Generation data sets do not support or enhance the persistence strategy (in-place table maintenance), and they do not work particularly well with SQL. However, they do complement audit trails, which we just examined, so this is a convenient place to consider them.

Reference: Read more about generation data sets in *SAS 9.2 Language Reference: Concepts*: SAS Files Concepts: SAS Data Files: Understanding Generation Data Sets.

Consider again the three alternative strategies for recording progressive stages of refinement in your data: succession (a series of distinctly named tables), replacement (a series of like-named tables, each of which takes the place of its predecessor), and persistence (one table that is changed in place). Now suppose that you want to keep track of the history of your data. In other words, you want to be able to see not just the "latest and greatest," but also the steps that got you there.

If you are following the succession strategy, there is no problem (as long as you keep the whole sequence of tables). If you are following the persistence strategy, you can use audit trails. However, if you are using the replacement strategy, you are essentially hiding your tracks, and there is nothing in your data library to show how you got where you are.

Generation data sets essentially let you convert the replacement practice to the succession practice and automate the details (specifically, by providing a naming convention and applying that convention to successive versions of a table). Let's look at a demonstration.

After refreshing DEMOLIB.FIFTEENUPS to its initial state, we run this code:

```
DATA GDS_Demo(GENMAX=4);
SET demolib.fifteenups;
RUN;
```

This creates a new table (GDS_DEMO) and activates the generation data set feature for that table. The GENMAX parameter specifies the number of generations to be retained.

Tip: Generation data sets can also be managed with PROC DATASETS.

Now let's run this code to create a new version of GDS_DEMO, this time including only 15-year-olds:

```
DATA GDS_Demo;
SET demolib.fifteenups;
WHERE age=15;
RUN;
```

Notice that here we did not code the GENMAX data set option. That's only done once, at the start, to turn on the feature. Next, we use another DATA step to create another version of GDS_DEMO, with fewer columns:

```
DATA GDS_Demo;
SET GDS_Demo;
KEEP fname sex age;
RUN;
```

To see what we have, we can run three PROC PRINT steps.

```
PROC PRINT DATA=GDS_Demo(GENNUM=-2);
RUN;
PROC PRINT DATA=GDS_Demo(GENNUM=-1);
RUN;
PROC PRINT DATA=GDS_Demo;
RUN;
```

The GENNUM= data set option enables us to reference earlier versions. The convention is that –1 indicates the version immediately preceding the latest version, –2 indicates the version before that, and so on. Of course, omitting the option altogether calls for the latest version. So the first PROC PRINT displays:

```
Obs     FName      Sex     Age     Height     Weight

 1      Janet      F       15      62.5       112.5
 2      Mary       F       15      66.5       112.0
 3      Philip     M       16      72.0       150.0
 4      Ronald     M       15      67.0       133.0
 5      William    M       15      66.5       112.0
```

The log reports:

```
NOTE: There were 5 observations read from the data set
WORK.GDS_DEMO (gennum=1).
```

Notice that the GENNUM= data set option can accept relative generation numbers, but the NOTE indicates the corresponding absolute generation. The second PROC PRINT then gives us the following, because at that point we subsetted the observations to include 15-year-olds only:

```
Obs     FName      Sex     Age     Height     Weight

 1      Janet      F       15      62.5       112.5
 2      Mary       F       15      66.5       112.0
 3      Ronald     M       15      67.0       133.0
 4      William    M       15      66.5       112.0
```

In the log we see:

```
NOTE: There were 4 observations read from the data set
WORK.GDS_DEMO (gennum=2).
```

If we did not have the generation data sets feature in effect, the first two versions of GDS_DEMO would be gone at this point.

Finally, the last PROC PRINT generates:

```
Obs     FName      Sex     Age

 1      Janet      F       15
 2      Mary       F       15
 3      Ronald     M       15
 4      William    M       15
```

This reflects the elimination of some of the original variables. There was no generation reference in the PROC PRINT code, and there is no generation reference in the NOTE; it simply reads:

```
NOTE: There were 4 observations read from the data set
WORK.GDS_DEMO.
```

So, by default, the latest generation is used and the whole generation structure remains in the background.

Now we're ready to attempt the same exercise using SQL. PROC SQL itself has no way to invoke generation data sets, but because data set options are available within PROC SQL, there is no problem. We first run the %REFRESH_EXAMPLE macro to empty the DEMOLIB library and create a fresh copy of the FIFTEENUPS table. Then we process this statement to create GDS_DEMO and activate the generation data sets feature:

```
PROC SQL;
CREATE TABLE GDS_Demo (GENMAX=4) AS
SELECT       *
FROM         demolib.fifteenups
;
QUIT;
```

Then we replace GDS_DEMO once:

```
PROC SQL;
CREATE TABLE GDS_Demo AS
SELECT       *
FROM         demolib.fifteenups
WHERE        age=15
;
QUIT;
```

and again:

```
PROC SQL;
CREATE TABLE GDS_Demo AS
SELECT       fname, sex, age
FROM         GDS_Demo
;
QUIT;
```

These statements parallel exactly the successive DATA steps we used earlier to produce the successive versions of GDS_DEMO, and the results are indistinguishable. However, the last statement triggers this message in the log:

```
WARNING: This CREATE TABLE statement recursively references the
target table. A consequence of this is a possible data
integrity problem.
```

We saw this earlier (see Section 9.2). However, with generation data sets in effect, the two references in the code to GDS_Demo actually pertain to two different tables, so the warning seems particularly unfounded. However, you might prefer to avoid such usage in order to avoid the appearance of deficiency in your code.

9.8 Summary

In this chapter we've considered three different ways (succession, replacement, and persistence) for recording progressive refinements in a data table. We then focused on the one (persistence) that makes changes to a table in place, without breaking the continuity of its existence. Four different categories of change were discussed:

- data changes
- metadata changes
- structural changes
- feature additions and removals (such as the addition of indexes)

We also considered creation of a table as an event separate from population of a table, and illustrated several creation methods.

Chapter 10

Views

A view can be thought of as a virtual table. The distinctive essence of a view is that only instructions (and not data) are stored when the view is defined. The instructions are not applied to the data source until the view is referenced. Consequently, the source does not have to exist when the view is defined, but it must exist whenever the view is used. This is the diametric opposite of the situation for a table. The source data for a table obviously must exist when the table is created, but are not needed to support later use of the table.

Both the DATA step and PROC SQL are capable of constructing views. These two kinds of views are interoperable in use, meaning that SQL views can be read by DATA steps and by procedures other than PROC SQL, and that PROC SQL can read DATA step views. However, the two kinds of views have some distinctly different properties. In contrast, tables created using PROC SQL are virtually indistinguishable from those generated by other parts of SAS.

Reference: Read more about views in *SAS 9.2 Language Reference: Concepts*: SAS Files Concepts: SAS Views.

Before we demonstrate the creation and use of views, let's create a small table that can be used for most of the examples. The code is:

```
PROC SQL;
CREATE TABLE preteen AS
SELECT      name as FName,
            sex,
            age,
            height FORMAT=6.1,
            weight FORMAT=6.1
FROM        sashelp.class
WHERE       age<13
;
QUIT;
```

The table PRETEEN looks like Exhibit 10-1.

Exhibit 10-1 PRETEEN

Fname	Sex	Age	Height	Weight
James	M	12	57.3	83.0
Jane	F	12	59.8	84.5
John	M	12	59.0	99.5
Joyce	F	11	51.3	50.5
Louise	F	12	56.3	77.0
Robert	M	12	64.8	128.0
Thomas	M	11	57.5	85.0

10.1 Defining Views

Suppose we want a view that subsets the PRETEEN table in both dimensions (column and row), delivering only the rows for 11-year-olds and only the first three columns.

Such a view can be created in the DATA step by running this code:

```
DATA v_eleven / VIEW=v_eleven;
SET preteen;
WHERE age=11;
KEEP fname sex age;
RUN;
```

We can do nearly the same thing with PROC SQL if we submit:

```
PROC SQL;
CREATE VIEW  v_eleven_sql AS
SELECT       fname,
             sex,
             age
FROM         preteen
WHERE        age=11
;
QUIT;
```

Notice that we gave the PROC SQL view a distinct name. That permits it to coexist with the DATA step view as we examine and compare the behavior of the two.

10.2 Using Views

Let's see what happens when we reference these views in SAS code. To keep things simple, we start with PROC PRINT, using it to display the content of V_ELEVEN, our DATA step view:

```
PROC PRINT DATA=v_eleven;
RUN;
```

The log reports:

```
NOTE: There were 2 observations read from the data set
WORK.PRETEEN.
      WHERE age=11;

NOTE: There were 2 observations read from the data set
WORK.V_ELEVEN.
```

Notice that the notes report on both the direct processing of the view and the indirect processing of the underlying source (PRETEEN). Here is the output:

```
Obs     FName     Sex     Age

 1      Joyce      F       11
 2      Thomas     M       11
```

Next we let **PROC PRINT** work with the SQL view to demonstrate the interoperability:

```
  PROC PRINT DATA=v_eleven_sql;
  RUN;
```

The results are identical (except that the second note in the log of course refers to WORK.V_ELEVEN_SQL instead of WORK.V_ELEVEN).

Let's continue by exercising the two views with PROC SQL. First we display the data produced by the SQL view. The code is:

```
  PROC SQL;
  SELECT       *
  FROM         v_eleven_sql
  ;
  QUIT;
```

There are no log messages tracing the data back to the source. The output looks like this:

```
FName     Sex     Age
----------------------
Joyce      F       11
Thomas     M       11
```

To complete this set of doubly symmetric examples, we can use PROC SQL to refer to the DATA step view, as in:

```
PROC SQL;
SELECT         *
FROM           v_eleven
;
QUIT;
```

The output is unchanged, but the log reports:

```
NOTE: There were 2 observations read from the data set
WORK.PRETEEN.
      WHERE age=11;
```

So we've seen that the two types of views are interchangeable, at least when used as inputs. The messages that appear in the log vary, and reflect interaction between the views themselves and the contexts in which they are used.

Now let's demonstrate an important property of views: that they automatically reflect changes in the content of their sources. Having earlier defined our two views, we now add a row to the underlying table by running:

```
PROC SQL;
INSERT INTO   preteen
SET           fname='Roger', age=11, sex='M'
;
QUIT;
```

The new row refers to an 11-year-old, so it ought to be admitted into each of our views. We can check by submitting these SQL statements:

```
PROC SQL;
SELECT         *
FROM           v_eleven
;

SELECT         *
FROM           v_eleven_sql
;
QUIT;
```

This is the output from the first:

```
FName     Sex      Age
----------------------
Joyce     F         11
Thomas    M         11
Roger     M         11
```

The output from the second is identical. So we have confirmed that views are up-to-date, not "frozen."

10.3 Inspecting Views

Both the DATA step and PROC SQL offer DESCRIBE statements to expose the instructions that are embedded in views. However, you have to know what kind of view you have; there is no interoperability here.

To inspect a DATA step view, we need a special kind of DATA step, which looks like this:

```
DATA VIEW=v_eleven;
DESCRIBE;
RUN;
```

When we run it, we see this in the log:

```
NOTE: DATA step view WORK.V_ELEVEN is defined as:

DATA v_eleven / VIEW=v_eleven;
   SET preteen;
   WHERE age=11;
   KEEP fname sex age;
RUN;
```

Except for some minor cosmetic differences, this is the code we used earlier to create this view. However, if we ask the DATA step to show us the code behind a PROC SQL view, as with:

```
DATA VIEW=v_eleven_sql;
DESCRIBE;
RUN;
```

we get:

```
ERROR: The view WORK.V_ELEVEN_SQL is either corrupt or
not a DATA step view.
```

So we see that the two kinds of views are different and not equally acceptable to the DATA step's DESCRIBE statement.

Now let's move over to the SQL context and test the DESCRIBE VIEW statement there. This is a different statement with a different syntax. First, let's try it with an alien view (that is, a DATA step view). The code is:

```
PROC SQL;
DESCRIBE VIEW v_eleven;
QUIT;
```

and it too runs into trouble. The log indicates:

```
WARNING: The View WORK.V_ELEVEN is not an SQL View.
```

and no code is displayed. However, when we use the statement appropriately by pointing it at a PROC SQL view, as with:

```
PROC SQL;
DESCRIBE VIEW v_eleven_sql;
QUIT;
```

we see:

```
NOTE: SQL view WORK.V_ELEVEN_SQL is defined as:

        select fname, sex, age
          from PRETEEN
          where age=11;
```

Except for the different use of white space, this accurately reflects the code we used when we defined the view.

10.4 Changing a Table via a View

Under some circumstances, you can use a PROC SQL view as a vehicle to update the underlying table. For example, we can run:

```
PROC SQL;
INSERT INTO   v_eleven_sql
SET           fname='Laura', age=11, sex='F'
;
QUIT;
```

The log confirms the operation, stating:

```
NOTE: 1 row was inserted into WORK.V_ELEVEN_SQL.
```

This is a bit puzzling, since V_ELEVEN_SQL is a view and thus does not actually contain rows of data. So let's look at the underlying table (PRETEEN), or at least the rows for 11-year-olds. If we submit:

```
PROC SQL;
SELECT      *
FROM        preteen
WHERE       age=11
;
QUIT;
```

we get:

FName	Sex	Age	Height	Weight
Joyce	F	11	51.3	50.5
Thomas	M	11	57.5	85.0
Roger	M	11	.	.
Laura	F	11	.	.

So, actually, the new row (for Laura) was inserted in the PRETEEN table. Of course it consequently is reflected in views that use PRETEEN as their data source.

Reference: For specific information on the limitations of updatable views, see the *Base SAS 9.2 Procedures Guide*: Procedures: The SQL Procedure: Concepts: SQL Procedure: Updating PROC SQL and SAS/ACCESS Views.

We can try changing PRETEEN via the DATA step view by running this code:

```
PROC SQL;
INSERT INTO  v_eleven
SET          fname='Andrea', age=11, sex='F'
;
QUIT;
```

When we run this statement, we get only this message:

```
ERROR: Unable to open the read only view WORK.V_ELEVEN
for UPDATE.
```

That is because DATA step views are read-only and cannot be used as vehicles for updating their source tables.

The pattern holds if we use a DATA step to make the changes. It is still only the PROC SQL view that accepts them. To demonstrate, we first use a DATA step with an SQL view to attempt again to append the row for Andrea. The code is:

```
DATA v_eleven_sql;
fname='Andrea'; age=11; sex='F';
OUTPUT;
STOP;
MODIFY v_eleven_sql;
RUN;
```

The log reflects success by reporting:

```
NOTE: The data set WORK.V_ELEVEN_SQL has been updated.
There were 0 observations rewritten, 1 observations
added and 0 observations deleted.
```

If we then run this PROC PRINT step:

```
PROC PRINT DATA=preteen;
WHERE age=11;
RUN;
```

our output looks like this:

```
Obs     FName     Sex     Age     Height     Weight

  4     Joyce       F      11       51.3       50.5
  7     Thomas      M      11       57.5       85.0
  8     Roger       M      11        .          .
  9     Laura       F      11        .          .
 10     Andrea      F      11        .          .
```

However, if we try using a DATA step view as the immediate target of a DATA step modification, as with this code:

```
DATA v_eleven;
fname='Gary'; age=11; sex='M';
OUTPUT;
STOP;
MODIFY v_eleven;
RUN;
```

we once again trigger this error:

```
ERROR: Unable to open the read only view WORK.V_ELEVEN
for UPDATE.
```

So, we conclude that PROC SQL views can be used as conduits for updating the underlying tables, but DATA step views cannot. It does not matter whether PROC SQL is used to make the updates.

10.5 Managing Library References

In the examples we've presented thus far in this chapter, we've used only the WORK library. When permanent libraries enter the picture, additional issues arise. We conclude our examination of views by exploring how views interact with library references (librefs).

First we establish the library reference DEMOLIB like this:

```
LIBNAME demolib "c:\temp\demolib";
```

This particular code is for Windows and requires modification in other environments.

Next, we create two subsets of the CLASS table, one (containing observations pertaining to the boys) in the WORK library and the other (holding the girls' data) in our DEMOLIB library:

```
DATA work.subset
     demolib.subset;
SET sashelp.class(RENAME = (name=FName) );
KEEP fname sex age;
IF sex='M' THEN OUTPUT work.subset;
IF sex='F' THEN OUTPUT demolib.subset;
run;
```

Reference: Read more about libraries in *SAS 9.2 Language Reference: Concepts*: SAS Files Concepts: SAS Libraries.

Now we create, in the DEMOLIB library, a DATA step view:

```
DATA demolib.v_twelve / view=demolib.v_twelve;
SET subset;
WHERE age=12;
RUN;
```

Notice that although the view is in the DEMOLIB library, it does not explicitly reference, in its SET statement, any library. Rather, the SET statement includes only a one-part name, identifying a data set (named SUBSET in this example) without identifying the library in which it is stored. So default behavior is in effect. Let's see how that works. If we run this PROC SQL statement:

```
PROC SQL;
SELECT          *
FROM            demolib.v_twelve
;
QUIT;
```

we get this output:

```
FName      Sex       Age
---------------------
James      M          12
John       M          12
Robert     M          12
```

The display of the boys' data rather than the girls' data tells us that the view looked in the WORK library for the data set SUBSET. That's not surprising; it's the usual behavior we see in SAS when there is no explicit library reference.

Now let's repeat the exercise, but use a PROC SQL view instead of a DATA step view. First we define the view with this code:

```
PROC SQL;
CREATE VIEW   demolib.v_twelve_sql AS
SELECT        *
FROM          subset
WHERE         age=12
;
```

Then we use the view in this SELECT statement:

```
SELECT        *
FROM          demolib.v_twelve_sql
;
QUIT;
```

The output looks like this:

```
FName     Sex       Age
-----------------------
Jane      F          12
Louise    F          12
```

These are girls' names, which we stored in DEMOLIB.SUBSET (whereas WORK.SUBSET contains boys' names). So, we conclude that a PROC SQL view that points to a one-part table name looks for that table in its "home" library. In this respect PROC SQL views are unlike DATA step views (and other parts of SAS), which expect such tables to be in the WORK library.

We've now thoroughly explored the behavior of views that lack explicit library references and rely instead on default rules. We conclude this chapter by demonstrating a feature of PROC SQL views that comes into play when library references **are** explicit.

First, let's remove the DEMOLIB library from the environment. We can do that with this statement:

```
LIBNAME demolib CLEAR;
```

Remember that this merely removes the pointer that we need to use the library; it does not affect the content.

Even though the library reference is gone, we can create a view that cites it, as in this example:

```
PROC SQL;
CREATE VIEW  v_fourteen_sql AS
SELECT       *
FROM         demolib.subset
WHERE        age=14
;
QUIT;
```

Of course, the view can't be used unless DEMOLIB is in effect as a library reference pointing to a library containing a suitable table named SUBSET. Otherwise, if we try to use it, as with this statement:

```
PROC SQL;
SELECT       *
FROM         v_fourteen_sql
;
QUIT;
```

an error is triggered:

```
ERROR: Libname DEMOLIB is not assigned.
```

When a view includes a two-part name to reference a table, the library reference portion of that name must exist when the view is used. That's not at all surprising. We could of course make the view usable by submitting an appropriate LIBNAME statement to create the library reference in the environment (SAS session or batch job) that is using the view. For PROC SQL views (but not for DATA step views), there is another option: a view that embeds its own environment. The code looks like this:

```
PROC SQL;
CREATE VIEW  v_fourteen_sql AS
SELECT       *
FROM         demolib.subset
WHERE        age=14
USING LIBNAME
             demolib "c:\temp\demolib";
;
QUIT;
```

A USING LIBNAME clause must appear at the end of a CREATE VIEW statement, after the SELECT clause and its subordinates, and after even the ORDER BY clause (if there is one). It creates the connection between the view and its intended data source.

So now we can again try running:

```
PROC SQL;
SELECT          *
FROM            v_fourteen_sql
;
QUIT;
```

This time it works, and we see:

```
FName      Sex        Age
----------------------
Carol      F           14
Judy       F           14
```

The local libref created by the USING LIBNAME clause is completely separate from any global librefs that might be in effect. It exists only as long as the view is processing. We can demonstrate that by running:

```
PROC CONTENTS DATA=demolib.subset;
RUN;
```

Because of the temporary nature of the embedded library reference, this generates an error:

```
ERROR: Libname DEMOLIB is not assigned.
```

10.6 Summary

Like the DATA step, PROC SQL can be used to build views (virtual tables). These two view types are interchangeable when used as data sources. However, there are significant differences:

- Some PROC SQL views can be used as intermediaries to update underlying tables. DATA step views cannot be used for that purpose.

- PROC SQL views and DATA step views have different rules for locating their data sources when those sources are specified using one-part names.

- PROC SQL views, but not DATA step views, can have embedded library references.

Chapter 11

PROC SQL as a Report Generator

PROC SQL is used most often to prepare data for subsequent processing. That means that queries are usually placed within CREATE TABLE or CREATE VIEW statements, or that INTO clauses are used to load results into macro variables. However, PROC SQL can also be used as a report generator and has a few features that extend that usefulness.

To explore PROC SQL's reporting capabilities through examples, we need a test table. As usual, we derive one from SASHELP.CLASS by running this code:

```
PROC SQL;
CREATE TABLE fifteens AS
SELECT      name as FName LABEL='First Name',
            sex,
            age,
            height FORMAT=6.1,
            weight FORMAT=6.1
FROM        sashelp.class
WHERE       age=15
;
QUIT;
```

Exhibit 11-1 displays the result (FIFTEENS).

Exhibit 11-1 FIFTEENS

FName	Sex	Age	Height	Weight
Janet	F	15	62.5	112.5
Mary	F	15	66.5	112.0
Ronald	M	15	67.0	133.0
William	M	15	66.5	112.0

11.1 Simple Reports

Let's start with the most basic report, which results from this minimal SELECT statement:

```
PROC SQL;
SELECT      *
FROM        fifteens
;
QUIT;
```

This is the output:

```
First
Name        Sex        Age   Height   Weight
-------------------------------------------
Janet       F           15    62.5    112.5
Mary        F           15    66.5    112.0
Ronald      M           15    67.0    133.0
William     M           15    66.5    112.0
```

Notice that the label "First Name" is automatically used over the FNAME column, and that there are no row numbers. So if we invoke PROC PRINT with the LABEL and NOOBS options, as in:

```
PROC PRINT DATA=fifteens LABEL NOOBS;
RUN;
```

we get this very similar output:

```
First
Name        Sex      Age      Height     Weight

Janet       F         15       62.5      112.5
Mary        F         15       66.5      112.0
Ronald      M         15       67.0      133.0
William     M         15       66.5      112.0
```

Before going on to more complicated situations, let's consider a variation on this basic report. Perhaps we want row numbers, and we also want the report double-spaced. If we change our PROC PRINT code by removing the NOOBS option and adding the DOUBLE option, so that it looks like this:

```
PROC PRINT DATA=fifteens LABEL DOUBLE;
RUN;
```

the output looks like this:

```
            First
Obs         Name      Sex      Age      Height     Weight

 1          Janet      F        15       62.5      112.5

 2          Mary       F        15       66.5      112.0

 3          Ronald     M        15       67.0      133.0

 4          William    M        15       66.5      112.0
```

We can make parallel changes in the PROC SQL output by invoking PROC SQL's DOUBLE option as well as its NUMBER option. This causes the output to include row numbers very much like the observation numbers that **PROC PRINT** presents by default. Then our code might look like this:

```
PROC SQL DOUBLE NUMBER;
SELECT       *
FROM         fifteens
;
QUIT;
```

When we run it, we get:

```
      First
Row   Name      Sex       Age   Height   Weight
-------------------------------------------------
  1   Janet     F          15    62.5    112.5

  2   Mary      F          15    66.5    112.0

  3   Ronald    M          15    67.0    133.0

  4   William   M          15    66.5    112.0
```

In the simple reports we've just considered, the PROC SQL code has been little more, or less, complicated than corresponding PROC PRINT code. Usually, we've needed only to code appropriate options to override defaults.

11.2 Complex Reports

When reports get a bit more complicated, things get a little more difficult for PROC SQL. For example, suppose that we want a list of student names, with sex indicators and with a count at the bottom of the list. That's pretty easy with PROC PRINT; we just include the N option in the PROC statement to get the count. The code is:

```
PROC PRINT DATA=fifteens N LABEL;
ID fname;
VAR sex;
RUN;
```

When we run it, we get:

```
First
Name       Sex

Janet       F
Mary        F
Ronald      M
William     M

N = 4
```

To get similar output from PROC SQL, we need code like this:

```
PROC SQL;
SELECT      fname, sex
FROM        fifteens
UNION ALL
SELECT      '', ''
FROM        fifteens(obs=1)
UNION ALL
SELECT      'N = ' || put(count(*),6. -L), ''
FROM        fifteens
;
QUIT;
```

which generates this output:

```
First Name  Sex
--------------
Janet        F
Mary         F
Ronald       M
William      M

N = 4
```

The first of the three UNION-linked SELECT clauses is responsible for the body of the table, and the last SELECT clause is responsible for the count that appears at the bottom. The middle SELECT clause is there only to create the intervening blank line that serves to make the count more prominent. This sort of tricky technique becomes more and more necessary as you try to build more complicated reports with SQL.

Let's consider another example, one that presents only summary data and no underlying detail. Specifically, suppose you need a report showing minimum, average, and maximum values for HEIGHT and WEIGHT. It's pretty easy with PROC MEANS; you run:

```
PROC MEANS DATA=fifteens MIN MEAN MAX MAXDEC=1;
VAR height weight;
RUN;
```

and you get:

```
Variable       Minimum            Mean       Maximum
---------------------------------------------------
Height            62.5            65.6          67.0
Weight           112.0           117.4         133.0
---------------------------------------------------
```

On the other hand, PROC SQL requires code along these lines:

```
SELECT      'Height'      LABEL = 'Variable',
            MIN(height)  LABEL = 'Minimum' FORMAT=12.1,
            MEAN(height) LABEL = 'Mean'    FORMAT=12.1,
            MAX(height)  LABEL = 'Maximum' FORMAT=12.1
FROM        fifteens
UNION ALL
SELECT      'Weight',
            MIN(weight),
            MEAN(weight),
            MAX(weight)
FROM        fifteens
  ;
```

It's not too complicated, but it is verbose. The output looks like this:

```
Variable       Minimum            Mean       Maximum
---------------------------------------------------
Height            62.5            65.6          67.0
Weight           112.0           117.4         133.0
---------------------------------------------------
```

In addition to complexity and verbosity, there is another consideration. Multiple queries connected by UNION operators cause the SQL processor to read the source table multiple times. In contrast, procedures like PROC MEANS are designed to produce such reports efficiently.

11.3 Reports with Long Character Strings

Now let's consider the special problem that arises when a table contains character columns populated with long values, making it difficult to produce tidy, readable reports. To illustrate PROC SQL's strengths and weaknesses in this area, we must contrive an example. We run this code to expand the FIFTEENS table by introducing two new character columns:

```
PROC SQL;
ALTER TABLE   fifteens
ADD           Comments CHAR(60),
              Notes    CHAR(60)
;
UPDATE        fifteens
SET           comments =
                'William hopes to study at ' ||
                'William and Mary, as does Mary.',
              notes =
                'That makes it hard to ' ||
                'display a table compactly.'
WHERE         fname='William'
;
UPDATE        fifteens
SET           comments =
                'Mary, like William, aspires ' ||
                'to attend William and Mary.',
              notes =
                'This is a second long text ' ||
                ' field in this table.'
WHERE         fname='Mary'
;
QUIT;
```

To facilitate the example, we also narrow the space available by submitting:

```
OPTIONS LS=68;
```

Now our challenge is to find a legible and compact way to present this table. PROC PRINT does a reasonably good job with minimal customization. We can run:

```
PROC PRINT DATA=fifteens NOOBS;
ID fname;
VAR comments sex notes height weight;
RUN;
```

which generates:

```
FName                                     Comments

Janet
Mary      Mary, like William, aspires to attend William and Mary.
Ronald
William   William hopes to study at William and Mary, as does Mary.

FName     Sex                              Notes

Janet      F
Mary       F        This is a second long text  field in this table.
Ronald     M
William    M        That makes it hard to display a table compactly.

FName     Height    Weight

Janet      62.5      112.5
Mary       66.5      112.0
Ronald     67.0      133.0
William    66.5      112.0
```

Essentially, PROC PRINT divides the data into column groups, each of which can fit in the available space. The ID statement specifies that FNAME is to appear repeatedly, once in each column group. That provides a reference for the reader.

PROC SQL has nothing corresponding to the ID statement in PROC PRINT, and it basically takes a simpler and less satisfactory approach. If we submit:

```
PROC SQL;
RESET        DOUBLE
;
SELECT       fname,
             comments,
             sex,
             notes,
             height,
             weight
FROM         fifteens
;
RESET        NODOUBLE
;
QUIT;
```

we get:

```
First
Name
Comments                                                             Sex
Notes                                                               Height
Weight
--------------------------------------------------------------------------
Janet
                                                                     F
                                                                      62.5
  112.5

Mary
Mary, like William, aspires to attend William and Mary.              F
This is a second long text field in this table.                      66.5
  112.0

Ronald
                                                                     M
                                                                      67.0
  133.0

William
William hopes to study at William and Mary, as does Mary.            M
That makes it hard to display a table compactly.                     66.5
  112.0
```

The necessary wraparound is performed separately for the column headings and for each row of data. The result is a bit of a jumble. It would be worse without the DOUBLE option, which inserts blank lines between the logical rows.

Fortunately, PROC SQL has a feature, the FLOW option, that does a better job of wrapping long text strings by treating each column separately. So the code becomes:

```
PROC SQL;
RESET        FLOW DOUBLE
;
SELECT       fname,
             comments,
             sex,
             notes,
             height,
             weight
FROM         fifteens
;
RESET        NOFLOW NODOUBLE
;
QUIT;
```

and the output looks like this:

```
First
Name      Comments         Sex  Notes            Height  Weight
----------------------------------------------------------------
Janet                      F                       62.5   112.5

Mary      Mary, like       F    This is a second   66.5   112.0
          William, aspires      long text field
          to attend William     in this table.
          and Mary.

Ronald                     M                       67.0   133.0

William   William hopes to M    That makes it      66.5   112.0
          study at William      hard to display a
          and Mary, as          table compactly.
          does Mary.
```

How about non-SQL solutions? PROC PRINT has no FLOW option, but PROC REPORT does. We can run:

```
PROC REPORT DATA=fifteens NOWD;
COLUMN fname comments sex notes height weight;
 DEFINE        fname    / GROUP                   ;
 DEFINE        comments / DISPLAY WIDTH=16 FLOW;
 DEFINE        sex      / DISPLAY WIDTH=3         ;
 DEFINE        notes    / DISPLAY WIDTH=16 FLOW;
 DEFINE        height   / DISPLAY                 ;
 DEFINE        weight   / DISPLAY                 ;
 BREAK BEFORE fname     / SKIP                    ;
 RUN;
```

The output looks like this:

```
First
Name         Comments        Sex   Notes          Height   Weight

Janet                         F                    62.5     112.5

Mary         Mary, like       F    This is a       66.5     112.0
             William,              second long
             aspires to            text field in
             attend William        this table.
             and Mary.

Ronald                        M                    67.0     133.0

William      William hopes    M    That makes it   66.5     112.0
             to study at           hard to display
             William and           a table
             Mary, as does         compactly.
             Mary.
```

We see in the next section that the Output Delivery System gives us a different way to solve the problem of presenting long string values.

11.4 PROC SQL and the Output Delivery System

PROC SQL supports the SAS Output Delivery System (ODS). A thorough treatment of the features and applications of ODS is beyond the scope of this book, so we'll limit the discussion to one example.

We've been looking at long string values and the issue of wraparound in the context of the default LISTING destination. Other ODS destinations have features that can mitigate the problems without requiring any special effort. For example, we can try a simple PROC PRINT step but route the output to the HTML destination by submitting something like:

```
ODS LISTING CLOSE;
ODS HTML FILE="&path\htmldemo.html";
PROC PRINT DATA=fifteens LABEL NOOBS;
VAR fname comments sex notes height weight;
RUN;
ODS HTML CLOSE;
ODS LISTING;
```

The exact appearance of the output depends on the Web browser. In general, HTML rendering lets text flow as necessary to achieve a compact layout, so that it looks something like Exhibit 11-2.

Exhibit 11-2 FIFTEENS (HTML)

First Name	Comments	Sex	Notes	Height	Weight
Janet		F		62.5	112.5
Mary	Mary, like William, aspires to attend William and Mary.	F	This is a second long text field in this table.	66.5	112.0
Ronald		M		67.0	133.0
William	William hopes to study at William and Mary, as does Mary.	M	That makes it hard to display a table compactly.	66.5	112.0

We can also run a similarly simple PROC SQL query and route its output to the HTML destination. The code would look like this:

```
PROC SQL;
ODS LISTING CLOSE;
ODS HTML FILE="&path\htmldemo.html";
SELECT      fname,
            comments,
            sex,
            notes,
            height,
            weight
FROM        fifteens
;
ODS HTML CLOSE;
ODS LISTING;
QUIT;
```

The output looks much like what we saw when we used PROC PRINT (see Exhibit 11-2). That's because it's the browser, not SAS, that is doing the rendering.

11.5 Summary

PROC SQL at its most basic is a report generator, but its capabilities in that realm are rather limited compared to what other parts of SAS offer. However, PROC SQL does have a few handy options that enhance the presentation of data. PROC SQL supports the Output Delivery System (ODS), which makes the features of ODS, and of ODS destinations and related software, available to enhance the appearance of SQL output.

Chapter 12

Mixed Solutions

In the preceding chapters, we've seen many examples of PROC SQL code together with equivalent or near-equivalent non-SQL SAS code. Depending on the task, one might be simpler and more natural to use than the other; in some cases it might seem to be a toss-up.

When we turn from narrow tasks to more complicated problems, we often find that the solution is a series of subtasks or operations. One operation might be more easily done with SQL, whereas another might be a natural for the DATA step, or vice versa. Fortunately, the interoperability between PROC SQL and the rest of SAS makes it easy to craft mixed solutions, taking advantage of each tool's strengths. One can use a table (SAS data file) or a view (as explained in Chapter 10) to pass intermediate results between PROC SQL and other PROC or DATA steps.

> **Reference:** Read more about views in *SAS 9.2 Language Reference: Concepts*: SAS Files Concepts: SAS Views.

We use some extended examples in this chapter to identify some of the relative strengths and weaknesses of SQL and to show how mixed solutions can be developed.

12.1 Example: Schedule Matrix

Suppose you are given data on the arrival and departure times of several individuals attending an event. You are then asked to produce a report showing not only those times, but also a triangular half-matrix table showing the overlap in attendance of all of the various pairings of two individuals. There are two input tables, ARRIVALS (see Exhibit 12-1) and DEPARTURES (see Exhibit 12-2).

Exhibit 12-1 ARRIVALS

Name	Arrival
John	14:30
Paul	15:00
Ringo	15:30
George	16:00

Exhibit 12-2 DEPARTURES

Name	Departure
John	15:30
Paul	16:00
Ringo	17:30
George	18:00

The result should be a two-part table along these lines:

```
Individual Presence

                      George        John         Paul         Ringo

            16:00-18:00  14:30-15:30  15:00-16:00  15:30-17:30

Joint Presence

                      George        John         Paul

John              . . . . . .
Paul              . . . . . .  15:00-15:30
Ringo          16:00-17:30   . . . . . .  15:30-16:00
```

The upper portion of this report is just a rearrangement of the given data. The lower part is more complicated and requires comparisons across individuals. The three intervals shown in the triangle are in fact the overlaps of the respective periods of presence. The dot-filled gaps correspond to pairs of individuals whose schedules do not overlap.

It looks kind of complicated, but the solution can be developed if the problem is broken down into stages. The fact that each arrival (or departure) time has to be compared with each other arrival (or departure) time should immediately suggest the use of a many-to-many join. A DATA step merge could of course serve to derive the upper part of the report, but would not readily facilitate the cross comparisons. So we want to start with a PROC SQL statement along these lines:

```
PROC SQL;
SELECT      *
FROM        Arrivals
            CROSS JOIN
            Departures
    ;
QUIT;
```

The result is:

```
Name        Arrival  Name       Departure
------------------------------------------
John         14:30   John         15:30
John         14:30   Paul         16:00
John         14:30   Ringo        17:30
John         14:30   George       18:00
Paul         15:00   John         15:30
Paul         15:00   Paul         16:00
Paul         15:00   Ringo        17:30
Paul         15:00   George       18:00
Ringo        15:30   John         15:30
Ringo        15:30   Paul         16:00
Ringo        15:30   Ringo        17:30
Ringo        15:30   George       18:00
George       16:00   John         15:30
George       16:00   Paul         16:00
George       16:00   Ringo        17:30
George       16:00   George       18:00
```

We could (after renaming one or both of the NAME columns) store these results in a table and turn to non-SQL tools to complete the derivation of the intervals. However, SQL is actually well-suited to that task, so we can instead run this more elaborate SQL statement:

```
PROC SQL;
CREATE VIEW  Pairs AS
SELECT           CASE WHEN arr.name > dep.name
                     THEN arr.name
                     ELSE dep.name
                     END AS UpDown_Name
             , CASE WHEN arr.name > dep.name
                     THEN dep.name
                     ELSE arr.name
                     END AS Across_Name
             , MAX(arrival)   AS LoTime FORMAT=TIME5.
             , MIN(departure) AS HiTime FORMAT=TIME5.
FROM         Arrivals   AS arr
             CROSS JOIN
             Departures AS dep
GROUP BY     UpDown_Name, Across_Name
ORDER BY     UpDown_Name, Across_Name
;
QUIT;
```

Notice that the CROSS JOIN specified in the FROM clause remains from the simple query shown earlier. We won't go through the rationale behind the rest of the code; just note that the statement is built up from elements we've seen in earlier chapters. The result (PAIRS) is presented in Exhibit 12-3.

Exhibit 12-3 PAIRS

UpDown_Name	Across_Name	LoTime	HiTime
George	George	16:00	18:00
John	George	16:00	15:30
John	John	14:30	15:30
Paul	George	16:00	16:00
Paul	John	15:00	15:30
Paul	Paul	15:00	16:00
Ringo	George	16:00	17:30
Ringo	John	15:30	15:30
Ringo	Paul	15:30	16:00
Ringo	Ringo	15:30	17:30

This is essentially the information needed for the report. The rows in which the two names are the same are to go in the upper segment, and the rows with differing names in the lower segment. The DATA step, and particularly the PUT statement, has features that provide a lot of control over layout, and SQL has little to match these features. So we turn to a DATA step to finish the job. Here's the code:

```
DATA _NULL_;
FILE PRINT N=PS;
SET Pairs;
BY updown_name;
IF FIRST.updown_name THEN DO;
   updown + 1;
   across = 0;
   END;
across + 1;
IF updown_name=across_name THEN
    PUT # 2 'Individual Presence'
        # 4 @(13*across) across_name $13.-R
        # 6 @;
```

```
ELSE PUT # 8 'Joint Presence'
        #10 @(13*across) across_name $13.-R
        #(updown+10) @1 updown_name @;
PUT          @(13*across) +3 @;
IF HiTime > LoTime THEN
    PUT LoTime +(-1) '-' HiTime;
ELSE PUT '. . . . . .';
RUN;
```

We won't go through all of the logic. The essential point is that we would be hard pressed to achieve such layout control using PROC SQL, so we take advantage of the DATA step's strength in this area. Here is the output:

```
Individual Presence

                  George        John         Paul        Ringo

          16:00-18:00  14:30-15:30  15:00-16:00  15:30-17:30

Joint Presence

                  George        John         Paul

John              . . . . . .
Paul              . . . . . .  15:00-15:30
Ringo             16:00-17:30  . . . . . .  15:30-16:00
```

12.2 Example: Identifying Spikes in a Series

Here is the scenario: We are given a one-column table containing a series of numeric readings. Our assignment is to produce a report containing the series and flagging each "spike," defined as a value that either exceeds both adjacent values by more than 3 or is less than both adjacent values by more than 3.

To create a test table, we run this code:

```
DATA Readings;
DO _n_ = 1 to 10;
   Reading = FLOOR(100 * RANUNI(2468) ) / 10;
   OUTPUT;
   END;
RUN;
```

The output (READINGS) is shown in Exhibit 12-4.

Exhibit 12-4 READINGS

Reading
4.8
1.7
7.6
1.9
5.1
1.7
0.7
2.1
8.9
2.8

For each value in the series, comparisons have to be made with the preceding term and with the succeeding term. In a DATA step, the look-back comparison is pretty simple, but the look-ahead one is not. SQL, on the other hand, is symmetric in that regard; both comparisons would be rather easy to code if the table included explicit row numbers. In the absence of such row numbers, things are difficult, because SQL does not let us reference rows in terms of order or positional relationship to other rows. Unfortunately, our table does not provide row numbers, and PROC SQL does not have a straightforward and documented way to generate them.

However, it is very easy to create row numbers in a DATA step. This leads us to a mixed solution. We use a DATA step to number our rows, then use SQL joins based on these numbers to make the needed comparisons.

Here is the DATA step:

```
DATA Numbered;
RowNum + 1;
SET Readings;
RUN;
```

The resulting table (NUMBERED) looks like Exhibit 12-5.

Exhibit 12-5 NUMBERED

RowNum	Reading
1	4.8
2	1.7
3	7.6
4	1.9
5	5.1
6	1.7
7	0.7
8	2.1
9	8.9
10	2.8

The table is used three times in this SQL step:

```
PROC SQL;
CREATE TABLE flagged AS
SELECT     This.Reading,
           CASE WHEN N( Prev.Reading,
                        This.Reading,
                        Next.Reading ) < 3
                THEN ''
                WHEN    This.Reading >
                        Prev.Reading + 3 AND
                        This.Reading >
                        Next.Reading + 3
                THEN 'High'
                WHEN    This.Reading <
                        Prev.Reading - 3 AND
                        This.Reading <
                        Next.Reading - 3
                THEN 'Low'
                ELSE ''
                END AS Flag
```

```
FROM        Numbered AS This
            LEFT JOIN
            Numbered AS Prev
            ON (This.RowNum - 1) = Prev.RowNum
            LEFT JOIN
            Numbered AS Next
            ON (This.RowNum + 1) = Next.RowNum
  ;
QUIT;
```

Let's consider the FROM clause first. It consists of two chained left joins. In such a construct, the first join is performed and its results enter the second join. Left joins are used here so that the results include rows for the first and last values in the series. These values do not have both look-ahead and look-back matches and would therefore fall out of an inner join.

Once the joins are completed by the SQL processor, each term in the series has the preceding and following terms available for comparison. The CASE expression uses them to derive the flags. The first branch in the CASE expression handles the first and last rows, which lack, respectively, look-back and look-ahead values. The second branch detects upward spikes (those where adjacent values are lower). Symmetrically, the third branch detects downward spikes (those where adjacent values are higher). The result (FLAGGED) is seen in Exhibit 12-6.

Exhibit 12-6 FLAGGED

Reading	Flag
4.8	
1.7	Down
7.6	Up
1.9	Down
5.1	Up
1.7	
0.7	
2.1	
8.9	Up
2.8	

12.3 Example: Using PROC TRANSPOSE to Normalize

We noted earlier (see Section 1.1) that SQL is very much designed to work with normalized data. Later we saw that macro code and macro variables provide workarounds, making it possible to get results from denormalized data without great difficulty; see Sections 8.1 and 8.2. These workarounds, however, are a second-best solution. If you intend to process your data with SQL, it's preferable to maintain the data in a normalized structure. So, let's revisit our earlier example and show how to do that.

To begin, let's re-create the denormalized table we used earlier:

```
DATA wide;
INPUT ID $ Estimated Net Gross Adjusted;
CARDS;
A 11 12 13 14
B 21 22 23 24
;
```

The result is shown in Exhibit 12-7.

Exhibit 12-7 WIDE

ID	Estimated	Net	Gross	Adjusted
A	11	12	13	14
B	21	22	23	24

The task, as before, is to sum each of the four numeric columns. When we did that earlier, the approach was to take the table in the form provided and directly produce a one-row, four-column table with the sums (see Exhibit 8-5). The code was a bit intricate. This time we'll preprocess the data and create a normalized structure that is much more suitable for use with SQL. Then we'll see that the code to compute sums becomes much simpler.

The SAS tool intended for such restructuring is the TRANSPOSE procedure. Here is code to normalize our data:

```
PROC TRANSPOSE DATA=wide
 OUT=long(rename = (_name_=Item col1=Value) );
BY ID;
RUN;
```

The new table is presented in Exhibit 12-8.

Exhibit 12-8 LONG (as produced by PROC TRANSPOSE)

ID	Item	Value
A	Estimated	11
A	Net	12
A	Gross	13
A	Adjusted	14
B	Estimated	21
B	Net	22
B	Gross	23
B	Adjusted	24

Can we accomplish this using PROC SQL? Yes, but we would face, again, the issue that SQL does not support arrays, so that analogous processing of multiple columns entails verbose and voluminous code. We can start by developing a simple query to handle just one of the numeric columns:

```
PROC SQL;
SELECT      ID, 'Estimated' AS Item, Estimated AS Value
FROM        wide
;
QUIT;
```

The result is:

```
ID          Item          Value
---------------------------------
A           Estimated        11
B           Estimated        21
```

Using this as a model, we can use similar queries for the other numeric columns, chain them together with UNION operators, and feed the results to a CREATE TABLE statement. Here is the code:

```
PROC SQL;
CREATE TABLE long AS
SELECT      ID, 'Estimated' AS Item, Estimated AS Value
FROM        wide
UNION ALL
SELECT      ID, 'Net'              , Net
FROM        wide
UNION ALL
SELECT      ID, 'Gross'            , Gross
FROM        wide
UNION ALL
SELECT      ID, 'Adjusted'         , Adjusted
FROM        wide
;
QUIT;
```

Contrast this with the simple and terse PROC TRANSPOSE code.

The output is shown in Exhibit 12-9. It is the same as that generated using PROC TRANSPOSE, except for the order of the rows. That is fine, since we are building it for use by PROC SQL, and PROC SQL is not sensitive to row ordering. That's why we did not specify an ORDER BY clause.

Exhibit 12-9 LONG (as produced by PROC SQL)

ID	Item	Value
A	Estimated	11
B	Estimated	21
A	Net	12
B	Net	22
A	Gross	13
B	Gross	23
A	Adjusted	14
B	Adjusted	24

Now we are ready to turn to the actual task. In terms of the restructured table, that task is to calculate sums of VALUE for each level of ITEM. As promised, the code is pretty simple:

```
PROC SQL;
CREATE TABLE verticalsums AS
SELECT     item, SUM(value) as Sum
FROM       long
GROUP BY   item
;
QUIT;
```

It does not matter which version of LONG is used, since they only differ in row order. The result can be seen in Exhibit 12-10.

Exhibit 12-10 VERTICALSUMS (alphabetical order)

Item	Sum
Adjusted	38
Estimated	32
Gross	36
Net	34

The computations are correct. The only hint of a potential problem is that the four ITEM strings appear in alphabetical order, an artifact of the processing triggered by the GROUP BY clause. They are not in what we presume to be the preferred order of the original table (ESTIMATED first, ADJUSTED last, NET before GROSS). Of course, as we've been saying, row order does not matter to SQL, so to the extent that this table is to be used in subsequent SQL processing, it's not a problem. However, at some point it might be necessary to produce a report in which it is desirable to use the preferred order for ITEM. Typically, that's a job for the ORDER BY clause, but we have no column that expresses the preferred order and is thus suitable for use in an ORDER BY clause. That could be a problem.

One solution is to exploit DICTIONARY tables to get an explicit indicator of the original column ordering. Specifically, we want the VARNUM column from DICTIONARY.COLUMNS. Here is the code:

```
PROC SQL;
CREATE TABLE numbered AS
SELECT      id, varnum, item, value
FROM        long
            INNER JOIN
            ( SELECT name, varnum
              FROM   dictionary.columns
              WHERE  libname='WORK' AND
                     memname='WIDE'
            )
            ON      name=item
;
QUIT;
```

The result (called NUMBERED) is reflected in Exhibit 12-11. It is superior to its ancestor (LONG) because it includes the VARNUM column expressing the order of the columns in WIDE. The VARNUM values range from 2 to 5 (not 1 to 4) because WIDE began with a column (ID) that did not become an ITEM level. Because the intended usage of VARNUM is ordinal rather than cardinal, that is not a problem.

Exhibit 12-11 NUMBERED

ID	varnum	Item	Value
A	2	Estimated	11
B	2	Estimated	21
A	3	Net	12
B	3	Net	22
A	4	Gross	13
B	4	Gross	23
A	5	Adjusted	14
B	5	Adjusted	24

Now we can compute the sums and control the order in which they are recorded. Here is the code:

```
PROC SQL;
CREATE TABLE verticalsums AS
SELECT      varnum, item, SUM(value) as Sum
FROM        numbered
GROUP BY    varnum, item
ORDER BY    varnum
;
QUIT;
```

The result is shown in Exhibit 12-12.

Exhibit 12-12 VERTICALSUMS (prescribed order)

varnum	Item	Sum
2	Estimated	32
3	Net	34
4	Gross	36
5	Adjusted	38

To conclude this example, we close the loop by putting the vector of totals back into a denormalized table structure. That's not appropriate for SQL processing, but it might be needed for reporting purposes. Here is the code:

```
PROC TRANSPOSE DATA=verticalsums
 OUT=horizontalsums(drop = _name_);
ID item;
VAR sum;
RUN;
```

If it's run against the version of VERTICALSUMS built with the ORDER BY clause, the original column order is reproduced, and the result looks like Exhibit 12-13.

Exhibit 12-13 HORIZONTALSUMS

Estimated	Net	Gross	Adjusted
32	34	36	38

This final reshaping could also have been done with PROC SQL, but the code is lengthy and rather convoluted.

12.4 Summary

SQL has strengths and weaknesses in comparison to non-SQL SAS tools. Sometimes it is advantageous to use the two together to solve a problem, drawing on the strengths of each tool.

Chapter **13**

Performance Tuning

Behind the scenes, the SQL processor looks at each PROC SQL statement and attempts to devise an efficient way to perform the required work. The component of the processor that does this planning is known as the optimizer. You, the programmer, can sometimes improve the outcome of the optimization and thus speed up the processing. There are basically two ways of doing this:

- providing resources (such as memory or indexes)
- coding in ways that give the optimizer "traction" or that compensate for its limitations

Performance tuning is a huge topic, so a thorough treatment is far beyond the scope of this book. We just present a couple of illustrative examples here.

13.1 Resource Example: The Effect of an Index

We start by constructing a table of 20 million rows, each containing a serial number and a random integer:

```
DATA myRandoms;
DO sernum = 1 TO 2E7;
    myRandom = ROUND(1E6*RANUNI(1) );
    OUTPUT;
    END;
RUN;
```

Next, we run this SQL code to find out how many rows match an arbitrary integer:

```
PROC SQL STIMER;
SELECT       COUNT(*)
INTO         : count654321
FROM         myRandoms
WHERE        myRandom = 654321
;
%PUT &count654321 rows counted.;
QUIT;
```

Note that the STIMER option has been invoked. This tells SAS to report on time usage for each statement, instead of reporting only at the conclusion of the PROC SQL step. This option is in effect for all of the examples in this chapter.

Here is the result of the query:

```
Occurrences
 of 654321
----------
        22
```

The issue is the amount of time required to locate these 22 rows among the original 20 million. The log tells us:

```
        real time        4.49 seconds
        cpu time         3.15 seconds
```

Now, we build an index:

```
PROC SQL STIMER;
CREATE INDEX myRandom ON myRandoms;
QUIT;
```

Of course this too requires time, quite a bit as it turns out. According to the log:

```
    real time          1:33.17
    cpu time           1:11.59
```

However, when we rerun the code to select the rows with the value 654321, we get the same result (22 rows), but the times are reduced:

```
    real time          1.81 seconds
    cpu time           0.02 seconds
```

The CPU time has declined from more than three seconds to a fiftieth of a second. That's because instead of reading and examining all 20 million rows, the computer was able to use the index to directly locate and count the matching rows. The 19,999,978 other rows were bypassed.

However, that time savings is just a small fraction of the time it took to build the index. We would have to use the index for many more tasks in order to recoup the investment made in creating it. So it's not a given that indexing is advantageous. In some situations the better strategy is to skip index creation and allow the slower sequential processing.

13.2 Code Example: The Advantage of Equijoins

The scenario: You have a table with people's names, phone numbers, and e-mail addresses. There is some duplication, and also inconsistency in how the names are recorded (e.g., nicknames versus formal names). Phone numbers and e-mail addresses are easier to standardize, and that's already been done. The present task is to detect possible duplicates by finding possible aliases: pairs of observations where either phone numbers or e-mail addresses (or both) match, but where names do not match.

Here's a test data generator:

```
DATA roster;
DO i = 1 TO 3e4; DROP i;
   name = i;
   phone = i + 0.1;
   email = i + 0.2;
   OUTPUT;
   IF RANUNI(111)>0.8  THEN name = name + 0.01;
   IF RANUNI(111)>0.8  THEN OUTPUT;
   END;
RUN;
```

Note that the two separate RANUNI calls in the two IF statements give rise to some cases where observations are duplicated exactly, and to other cases where the extra observations display variation in NAME values.

The data are not realistic in that all of the columns are numeric and the values don't look like names, phone numbers, or e-mail addresses. The table is suitable nevertheless for modeling and demonstration purposes. There are around 36,000 observations generated.

To start, let's construct a query that finds pairs of names for which the telephone number is the same. Here's the code:

```
PROC SQL STIMER;
CREATE TABLE same_phone AS
SELECT      DISTINCT roster.name,
                     copy.name AS diff_name
FROM        roster JOIN roster AS copy
ON          roster.phone=copy.phone
WHERE       roster.name LT copy.name
;
```

This is a self-join in which each row of ROSTER is a candidate to be matched with each other row of ROSTER. Since ROSTER has more than 30,000 rows, the theoretical size of this join is in the neighborhood of a billion rows. So let's see how long it took to run. The log reports:

```
real time           0.51 seconds
cpu time            0.13 seconds
```

This speedy performance is due to optimization. The SQL processor did not evaluate each possible pair of rows. Instead, it read each row, took the phone number, and used that phone number to more or less directly locate any matching rows.

Now suppose that we want to expand the results by considering pairs of names where either the phone numbers or the e-mail addresses (or possibly both) match. All we have to do is code the condition for matching e-mail addresses and connect it to the phone-number condition with an OR (because we are requiring only one of the items to match, not both). The query becomes:

```
PROC SQL STIMER;
CREATE TABLE slow AS
SELECT        DISTINCT roster.name,
                       copy.name AS diff_name
FROM          roster JOIN roster AS copy
ON            roster.phone=copy.phone OR
              roster.email=copy.email
WHERE         roster.name LT copy.name;
QUIT;
```

The log shows:

```
NOTE: The execution of this query involves performing
one or more Cartesian product joins that can not be
optimized.
```

As to time, we see:

```
    real time           3:10.75
    cpu time            3:07.12
```

Because of the OR in the ON clause, the SQL processor could not optimize the evaluation by using information from one side of the join to directly locate the appropriate rows from the other side. Instead it had to examine all of the potential name pairs, and there are more than a billion of those. The code works, but it's slow. Even worse, we can expect the time to increase as a quadratic function of table size.

However, we can separate the query into two parts, one for each of the join conditions, and use the UNION operator to combine the results:

```
PROC SQL STIMER;
CREATE TABLE fast AS
SELECT          roster.name,
                copy.name AS diff_name
FROM            roster JOIN roster AS copy
ON              roster.phone=copy.phone
WHERE           roster.name NE copy.name
UNION
SELECT          roster.name,
                copy.name AS diff_name
FROM            roster JOIN roster AS copy
ON              roster.email=copy.email
WHERE           roster.name NE copy.name;
QUIT;
```

The DISTINCT specification can be omitted now because the UNION operator has the same effect. Logically, the two versions are equivalent, and they produce the same results. However, the performance is dramatically improved. For the version using the UNION operator, we see:

real time	0.21 seconds
cpu time	0.20 seconds

The time required is now less than a second, a tiny fraction of what it was when we ran the form using the OR operator. That's because the SQL processor was able to optimize, separately, each of the UNION operator's two operands. Instead of examining a billion rows, the computer searched over merely tens of thousands of rows twice, and then combined those results.

13.3 Summary

PROC SQL has to take the specifications coded in a statement and devise a process or series of processes that will deliver the correct results. It attempts to optimize this processing in terms of speed. However, in some situations you can facilitate this by providing resources such as indexes or by coding in a way that assists the optimizer.

Chapter 14

Documentation Roadmap

The SAS documentation is extremely thorough, and it is organized so that it is often easy to locate the particular information you need. For example, consider the complex PUT statement found in the example presented in Section 12.1. The features used are explained in the section of the documentation covering the PUT statement, and the location of that section is pretty straightforward (*SAS 9.2 Language Reference: Dictionary*: Dictionary of Language Elements: Statements). Similarly, documentation for Base SAS procedures is found in the *Base SAS 9.2 Procedures Guide*, with the material organized in alphabetical order by procedure name.

SQL is something of an exception when it comes to ease of finding things in the documentation. That's because SQL is more than another SAS procedure; it is a language, and one that was not designed by SAS developers. As a consequence, the documentation for PROC SQL is different from what is familiar to most SAS users. This chapter offers some commentary intended to help you understand how the PROC SQL documentation is organized.

14.1   Where to Start?

If you look at the SAS 9.2 product documentation in the Knowledge Base (at http://support.sas.com/documentation/onlinedoc/base/), you will see the following documents among those listed under "Base 9.2 SAS":

- *Base SAS 9.2 Procedures Guide*
- *SAS 9.2 SQL Procedure User's Guide*
- *SAS 9.2 SQL Query Window User's Guide*

Then, **within** the *Base SAS 9.2 Procedures Guide*, there is a chapter on the SQL procedure.

So what's the difference, and where do you turn? There are four options:

- *SAS 9.2 SQL Query Window User's Guide*
- *SAS 9.2 SQL Procedure User's Guide*
- "The SQL Procedure" chapter of the *Base SAS 9.2 Procedures Guide*
- None of the above

Note: At this point we shift our attention from the HTML (web pages) in the Knowledge Base to the printed (or PDF) versions. That's just for convenience; the content is the same whether you look at the HTML, PDF, hard copy, or the Help system installed with your copy of SAS. Excerpts from the printed documentation that appear in boxes in the following sections are (except where noted) direct quotes, although typography, indention, and the like have been altered for the sake of emphasis.

SAS 9.2 SQL Query Window User's Guide

This is a specialized document pertaining to a graphical interface that lets you generate SQL code by doing a lot of pointing and clicking. By today's standards, and compared to tools like SAS Enterprise Guide and SAS Data Integration Studio, the SQL Query Window is rather primitive. So, unless you have a very specific reason for working with the SQL Query Window, you probably want to look elsewhere for information about PROC SQL.

SAS 9.2 SQL Procedure User's Guide

To get a sense of what the *SAS 9.2 SQL Procedure User's Guide* is about, look at the table of contents:

Notice the clear progression from the simple and elementary in the early chapters to the complex and advanced in the later chapters. This is an instructional work, one that does a pretty good job of explaining SQL as it is implemented within SAS. However, it is not a complete reference and syntax guide.

Base SAS 9.2 Procedures Guide (SQL Chapter)

This is the location of the syntax reference for PROC SQL, and is the main focus of the remainder of this chapter. However, it does not resemble the documentation for most SAS procedures. It is much more modular and much more dependent on cross references.

Let's look at the relevant part of the table of contents from the front of the *Base SAS 9.2 Procedures Guide*:

We can see from the progression of the page numbers that a lot of the pages are in the syntax section and the component dictionary. When we look at the beginning of the chapter, a more detailed table of contents is available. There we can see that the syntax section presents the **statements** that constitute SQL as implemented by SAS. The PROC statement appears first, followed by the others in alphabetical order. To this extent, the structure conforms to the usual template for documenting SAS procedure syntax.

Among these statements, SELECT is very much the "800-pound gorilla." It is complex and versatile, and much used (either as a stand-alone statement or, more often, as a clause within another statement) in most SQL applications. Most of the other statements are relatively specialized (such as some pertaining to the Pass-Through Facility and some used only in applications that change existing tables), relatively simple (for example, the DROP statement), or even non-essential (for example, the VALIDATE statement). That is not to say that they are not useful or important. The point is that this part of the table of contents does not provide much differential guidance; many users find themselves delving into the SELECT documentation constantly and the rest of the statement explanations rarely.

Now let's turn to the sections of the component dictionary.

SQL Procedure Component Dictionary 1180

BETWEEN condition 1180
BTRIM function 1181
CALCULATED 1182
CASE expression 1182
COALESCE Function 1184
column-definition 1184
column-modifier 1186
column-name 1187
CONNECTION TO 1188
CONTAINS condition 1189
EXISTS condition 1189
IN condition 1190
IS condition 1190
joined-table 1191
LIKE condition 1201
LOWER function 1203
query-expression 1203
sql-expression 1210
SUBSTRING function 1217
summary-function 1218
table-expression 1225
UPPER function 1226

Words and phrases in uppercase are used literally in SQL code. Components that are entirely in lowercase are placeholders that serve only in the documentation; they are replaced by other elements in SQL code, according to menus and rules that appear in the documentation.

The component dictionary is something of an odd mixture in other respects as well. For the most part, it includes things **not** specific to a single context (statement, clause, or component) or that don't fit in anywhere else, but there are exceptions. On the one hand, the ORDER BY clause (which is multi-context, being usable in the SELECT, CREATE TABLE, and CREATE VIEW statements) is **not** in the table of contents. On the other hand, the joined-table item is used **only** in the FROM clause of the SELECT statement, and so might have been subsumed there. Many of the component dictionary items are specialized or are presented for completeness (for example, the UPPER function, which is just an SQL-standard alias for the SAS UPCASE function).

The important point here is that this dictionary is definitely **not** a complete presentation of important SQL components, many of which are submerged elsewhere, either in the documentation of statements or in the documentation of other components. For example, subqueries are covered in the sql-expression section.

In looking at the component dictionary, we see a fair amount of jargon. That is, there are terms like "sql-expression" that are not part of the language per se (and which are therefore in lowercase) but that are used in the documentation to categorize and classify language constructs. Some of these terms are rather similar, which can make things more confusing. It is likely that many highly competent SQL programmers would find it difficult to distinguish among the following:

- sql-expression
- table-expression
- query-expression

Such experts would know very well how to form and use these three elements, but the terminology is not in everyday use. We will revisit these three terms a couple of times before we are done.

It is the jargon and the extensive cross references that can make the SQL reference documentation presented in the *Base SAS 9.2 Procedures Guide* somewhat perplexing. The remainder of this chapter is devoted to tips and explanations intended to reduce the confusion.

None of the Above

Before we continue examining the SQL reference documentation, it is important to understand that a lot of code elements allowed within a PROC SQL step are not, strictly speaking, part of SQL. Thus they are not documented in the SQL chapter of the *Base SAS 9.2 Procedures Guide*. PROC SQL is part of SAS, and therefore can "borrow" a lot of SAS features. We can see this at two distinct levels:

- Nearly all SAS functions (but not CALL routines), formats, informats, and data set options can be used in appropriate contexts within SQL statements. However, the "SQL Procedure Component Dictionary" in the *Base SAS 9.2 Procedures Guide* covers only functions that are exclusive to SQL and not usable in the DATA step.
- As explained in Section 7.1, most if not all SAS global statements can be interspersed with the SQL statements within a PROC SQL step (that is, after the PROC SQL statement and before the QUIT statement).

These non-SQL language elements used within PROC SQL steps are documented primarily in the *SAS 9.2 Language Reference: Dictionary*, so that manual should be considered part of the SQL user's reference set.

14.2 Following Cross References

The PROC SQL reference documentation is characterized by an unusual amount of cross referencing. We can get a feel for this by starting with the centerpiece of SQL, the SELECT statement.

SELECT Statement

The syntax skeleton for the SELECT statement begins with:

> SELECT <DISTINCT> *object-item* <, …*object-item*>

Here we've encountered another bit of jargon, "object item." It's just a placeholder, and the documentation following the skeleton lists the possible substitutions:

> *object-item* is one of the following:
>
> - *
>   represents all columns in the tables or views that are listed in the FROM clause.
> - case-expression <AS *alias*>
>   derives a column from a CASE expression. See "CASE expression" on page 1182.
> - column-name <<AS> *alias*> <column-modifier <… column-modifier>>
>   names a single column. See "column-name" on page 1187 and "column-modifier" on page 1186.
> - sql-expression <AS *alias*> <column-modifier <… column-modifier>>
>   derives a column from an sql-expression. See "sql-expression" on page 1210 and "column-modifier" on page 1186.

Of these possibilities, "sql-expression" is the most vague, so its definition might be the one we are most likely to need. We can follow the cross reference to "sql-expression."

sql-expression

The definition tells us that sql-expression:

> produces a value from a sequence of operands and operators.

It then presents the syntax skeleton:

> *operand operator operand*

and states that:

> *operand* is one of the following:
>
> - a *constant*, which is a number or a quoted character string (or other special notation) that indicates a fixed value. Constants are also called *literals*. Constants are described in *SAS Language Reference: Dictionary*.
> - a column-name, which is described in "column-name" on page 1187.
> - a CASE expression, which is described in "CASE expression" on page 1182.
> - any supported SAS function. PROC SQL supports many of the functions available to the SAS DATA step. Some of the functions that aren't supported are the variable information functions, functions that work with arrays of data, and functions that operate on rows other than the current row. Other SQL databases support their own sets of functions. Functions are described in the *SAS Language Reference: Dictionary*.
> - any functions, except those with array elements, that are created with PROC FCMP.
> - the ANSI SQL functions COALESCE, BTRIM, LOWER, UPPER, and SUBSTRING.
> - a summary-function, which is described in "summary-function" on page 1218.
> - a query-expression, which is described in "query-expression" on page 1203.
> - the USER literal, which references the userid of the person who submitted the program. The userid that is returned is operating environment-dependent, but PROC SQL uses the same value that the &SYSJOBID macro variable has on the operating environment.

Basically, this is telling us that an sql-expression is much like an expression that is coded in a DATA step, involving constants, variables (here called columns), operators, and functions. CASE expressions and summary functions are notable SQL extensions to this vocabulary. Putting aside the extremely specialized ANSI SQL functions and USER literal, we are left with "query-expression," the least concrete of the permitted ingredients. So we follow its cross reference.

query-expression

The explanation of query-expression is that it:

> retrieves data from tables.

The documentation then offers several cross references for the whole concept:

See also:

- "table-expression" on page 1225,
- "Query Expressions (Subqueries)" on page 1213, and
- "In-Line Views" on page 1173

Then we see the syntax skeleton:

table-expression <*set-operator* table-expression> <...*set-operator* table-expression>

This is followed by a list of the components:

Arguments

- table-expression
  is described in "table-expression" on page 1225.
- *set-operator*
  is one of the following:
  - INTERSECT <CORRESPONDING> <ALL>
  - OUTER UNION <CORRESPONDING>
  - UNION <CORRESPONDING> <ALL>
  - EXCEPT <CORRESPONDING> <ALL>

It then goes on to explain in considerable detail these set operators. However, the basic ingredient here is "table-expression"; we follow its cross reference.

table-expression

The documentation tells us that a table-expression:

defines part or all of a query-expression.

It then offers a cross reference:

See also: "query-expression" on page 1203

But it's a cross reference **from** query-expression that brought us here, so we don't want to follow this one right back there. Reading on, we get the syntax skeleton for a table-expression:

SELECT <DISTINCT> *object-item*<, ... *object-item*>

<INTO :*macro-variable-specification* <, ... :*macro-variable-specification*>>

FROM *from-list*

<WHERE sql-expression>

<GROUP BY *group-by-item* <, ... *group-by-item*>>

<HAVING sql-expression>

This is the nucleus of SQL. Yet the clauses are not explained below the skeleton. Moreover, it turns out that very few cross references from elsewhere in the documentation point here. That's a bit strange for the nucleus of the language.

Tip: The syntax for table-expression deliberately and appropriately excludes the ORDER BY clause, which is often taught as being part of the nucleus but is actually more of a post-processing specification.

There is one more cross reference:

See "SELECT Statement" on page 1166 for complete information on the SELECT statement.

This tells us where the details are. However, we **started** this tour with the SELECT statement documentation. Now we've come around in a circle. The problem is not as serious as it seems, and does not indicate any flaw in the manual. It is, in essence, due to the nestability of SQL, and, in particular, to the use of inline views (see Section 3.3) and subqueries (see Chapter 5). When we saw the admissibility of a query-expression as an operand within an sql-expression, we were in fact reading about subqueries. The lesson is that, in the PROC SQL reference documentation, you cannot mechanistically follow cross references from entities to their constituent entities and expect the process to end by leading you to nothing but primitives.

14.3 The Three Expressions Revisited

Our circular tour touched on the three types of expressions (sql-expression, table-expression, and query-expression). You might have found the explanations to be a bit nuanced. Let's consider the three again, using an example. Once again, our starting point is the SELECT statement.

SELECT Statement

Here is a valid (though not very useful) SELECT statement:

```
PROC SQL;
SELECT      name, age
FROM        sashelp.class
WHERE       sex='F'
UNION
SELECT      name, age + 1
FROM        sashelp.class
ORDER BY    age
;
QUIT;
```

query-expression

A query-expression is either a table-expression or two or more table-expressions connected by set operators. To illustrate, the query-expression is underlined:

```
PROC SQL;
SELECT      name, age
FROM        sashelp.class
WHERE       sex='F'
UNION
SELECT      name, age + 1
FROM        sashelp.class
ORDER BY    age
;
QUIT;
```

It's a little odd to see that the statement does not "sandwich" its major ingredient (the query-expression). Instead, the statement **begins** with the query-expression, followed by the ORDER BY clause and the terminating semicolon. The documentation does not exactly recognize this structure, and it tends to blur the distinction between the SELECT statement and its major ingredient, the query-expression.

table-expression

A table-expression is essentially a SELECT clause (not statement), with its subordinate clauses. The two table-expressions are underlined here:

```
PROC SQL;
SELECT          name, age
FROM            sashelp.class
WHERE           sex='F'
UNION
SELECT          name, age + 1
FROM            sashelp.class
ORDER BY        age
;
QUIT;
```

sql-expression

An sql-expression is a scalar expression (one which evaluates to a single value, and not to multiple rows or columns). In that way it's a lot like a formula or expression you might code in a DATA step. An sql-expression can incorporate subqueries, but each such subquery must ultimately evaluate to a scalar. Of course, an sql-expression that applies to a data source comprising multiple rows is evaluated repeatedly, once for each row. In that sense it can give rise to a vector, even though each evaluation generates a scalar result.

The sql-expressions are underlined in this presentation:

```
PROC SQL;
SELECT          name, age
FROM            sashelp.class
WHERE           sex='F'
UNION
SELECT          name, age + 1
FROM            sashelp.class
ORDER BY        age
;
QUIT;
```

The sql-expression is a ubiquitous construct. Here we see one as an object-item (that is, in a SELECT list) and one in a WHERE clause. Sql-expressions are also permitted in the ON clause (part of a join specification), the GROUP BY clause, the HAVING clause, and the ORDER BY clause.

14.4 Could It Be More Logical?

We noted earlier a bit of oddity in the documentation of the table-expression. Even though the table-expression (that is, the SELECT/FROM/WHERE/GROUP

BY/HAVING sequence) is the nucleus of SQL, its documentation is only sketchy, with the details being located instead under the SELECT statement. This is a reasonable arrangement in that it conforms to what most people probably expect. It is not, however, the most logical arrangement.

Before continuing this discussion of the documentation structure, let's consider another question: What are the devices available to make SQL results available outside PROC SQL? There are basically four such vehicles.

- Tables. SQL results can be placed in tables, which in turn can be used by other parts of the SAS System. This can be done with either the CREATE TABLE statement (see Section 2.3) or the INSERT statement (see Section 9.4).

- Views. The CREATE VIEW statement (see Section 10.1) can be used to make SQL results available to other parts of the SAS System when such results are needed.

- Output Delivery System. With or without the use of explicit ODS code, a stand-alone or "naked" SELECT statement (that is, a statement that begins with the keyword "SELECT") ordinarily sends its results to the Output Delivery System.

- Macro variables. A stand-alone SELECT statement that includes an INTO clause and that does not involve set operators (that is, which incorporates a simple table-expression and not a general query-expression) populates one or more macro variables with its results (see Section 8.2).

The preceding example:

```
PROC SQL;
SELECT      name, age
FROM        sashelp.class
WHERE       sex='F'
UNION
SELECT      name, age + 1
FROM        sashelp.class
ORDER BY    age
;
QUIT;
```

sends its results to the Output Delivery System. In the absence of any explicit ODS coding, and assuming that the code is run via the SAS Display Manager, that simply means that the results appear in the Output window.

Now suppose that instead of seeing the results, we want them in a SAS data set to be used as input to some SAS procedure. We can accomplish this by composing a CREATE TABLE statement to swallow our query:

```
PROC SQL;
CREATE TABLE mytable AS
SELECT      name, age
FROM        sashelp.class
WHERE       sex='F'
UNION
SELECT      name, age + 1
FROM        sashelp.class
ORDER BY    age
;
QUIT;
```

Each of these statements evaluates the same query-expression, then sorts the rows in the result set, if necessary, to conform to the ORDER BY specification. Instead of sending the ordered result set to ODS, the CREATE TABLE statement sends it to a SAS data set. Even though the CREATE TABLE statement appears to be a derivative of the SELECT statement, the two are, functionally, more like peers with a common core (the query-expression).

The syntax skeleton for the form of CREATE TABLE statement we are using here is:

CREATE TABLE *table-name* AS query-expression

<ORDER BY *order-by-item*<, ... *order-by-item*>>;

This suggests that the syntax skeleton for the stand-alone SELECT statement (which can be thought of as the "Feed ODS and/or Populate Macro Variables" statement) could be simply:

query-expression

<ORDER BY *order-by-item*<, ... *order-by-item*>>;

Note that this is **not** a quote from the manual.

The details of the SELECT/FROM/WHERE/GROUP BY/HAVING sequence could then be relocated to the section of the documentation that explains table-expressions.

14.5 Summary

SQL is very different from other SAS procedures, and as a consequence its documentation is organized differently. Here are some things to keep in mind when looking for answers about PROC SQL. We'll start with tips about **where** to look.

- Don't look for information in the *SAS 9.2 SQL Query Window User's Guide* unless you have a specific interest in that product (SQL Query Window).

- Use the *SAS 9.2 SQL Procedure User's Guide* to learn about SQL from explanations and examples.

- Consult the SQL chapter in the *Base SAS 9.2 Procedures Guide* to find the syntax rules for PROC SQL and to determine just what is allowed within each construct.

- Remember that the table of contents at the beginning of that chapter does not identify all of the major elements of SQL, but does point to a number of very minor elements.

- PROC SQL permits you to use many non-SQL SAS language elements (functions, data set options, global statements, and so on), so consider the *SAS 9.2 Language Reference: Dictionary* to be part of the documentation of PROC SQL.

The tips that follow all pertain to the syntax information in the *Base SAS 9.2 Procedures Guide*.

Try to learn the terminology and concepts used in the SQL syntax explanations, especially the three generic-sounding "expressions" (sql-expression, table-expression, and query-expression).

- An sql-expression is basically a scalar formula. Subqueries are components of sql-expressions, so the syntax documentation for subqueries is found in the sql-expression section.

- A table-expression is a SELECT clause (not statement) together with its required FROM clause and any of the optional subordinate clauses (INTO, WHERE, GROUP BY, or HAVING).

- A query-expression is either a single table-expression or multiple table-expressions connected using set operators. Because this is the only context in which set operators are used, they are documented in the query-expression section.

The modularity of SQL leads to a lot of cross referencing in the syntax documentation. The nestability provided by subqueries and inline views creates some circular paths through the cross referencing.

The syntax documentation for the nucleus of SQL (the SELECT clause with its subordinate FROM, INTO, WHERE, GROUP BY, and HAVING clauses) is found in the section for the SELECT statement, even though it logically belongs under the table-expression concept and is often used in CREATE TABLE and other statements.

Appendix A

SASHELP.CLASS Data Set

Many of the examples in the book use either the SASHELP.CLASS table or tables derived from it. Typically, that table is made available as part of your SAS software installation. If for some reason you don't have it, you can create the CLASS table by running this code:

```
PROC SQL;
CREATE TABLE mylib.class
          ( Name    char(8),
            Sex     char(1),
            Age     num,
            Height  num,
            Weight  num
          )
   ;
```

```
INSERT INTO  mylib.class
VALUES      ("Alfred" , "M", 14, 69.0, 112.5)
VALUES      ("Alice"  , "F", 13, 56.5,  84.0)
VALUES      ("Barbara", "F", 13, 65.3,  98.0)
VALUES      ("Carol"  , "F", 14, 62.8, 102.5)
VALUES      ("Henry"  , "M", 14, 63.5, 102.5)
VALUES      ("James"  , "M", 12, 57.3,  83.0)
VALUES      ("Jane"   , "F", 12, 59.8,  84.5)
VALUES      ("Janet"  , "F", 15, 62.5, 112.5)
VALUES      ("Jeffrey", "M", 13, 62.5,  84.0)
VALUES      ("John"   , "M", 12, 59.0,  99.5)
VALUES      ("Joyce"  , "F", 11, 51.3,  50.5)
VALUES      ("Judy"   , "F", 14, 64.3,  90.0)
VALUES      ("Louise" , "F", 12, 56.3,  77.0)
VALUES      ("Mary"   , "F", 15, 66.5, 112.0)
VALUES      ("Philip" , "M", 16, 72.0, 150.0)
VALUES      ("Robert" , "M", 12, 64.8, 128.0)
VALUES      ("Ronald" , "M", 15, 67.0, 133.0)
VALUES      ("Thomas" , "M", 11, 57.5,  85.0)
VALUES      ("William", "M", 15, 66.5, 112.0)
;
QUIT;
```

Notice that you must first code an appropriate LIBNAME statement. Then change the examples in the book to point to it rather than to SASHELP, which ought to be treated as read-only.

Appendix B

Online Resources

For additional information on this book, including example code, please visit support.sas.com/authors/schreier.html. Then, click Example Code and Data. To submit a comment, ask a question, or see other readers' questions, answers, and comments, please visit http://www.sascommunity.org/wiki/PROC_SQL_by_Example.

Index

ACCELERATE YOUR SAS® KNOWLEDGE WITH SAS BOOKS

Learn about our authors and their books, download free chapters, access example code and data, and more at **support.sas.com/authors**.

Browse our full catalog to find additional books that are just right for you at **support.sas.com/bookstore**.

Subscribe to our monthly e-newsletter to get the latest on new books, documentation, and tips—delivered to you—at **support.sas.com/sbr**.

Browse and search free SAS documentation sorted by release and by product at **support.sas.com/documentation**.

Email us: sasbook@sas.com
Call: 800-727-3228

THE POWER TO KNOW.®

CPSIA information can be obtained at www.ICGtesting.com
Printed in the USA
LVOW03s0752260714

395759LV00006B/35/P